Life-Saving Appliances

including **LSA** Code

2010 Edition

INTERNATIONAL
MARITIME
ORGANIZATION

London, 2010

Published in 2010
by the INTERNATIONAL MARITIME ORGANIZATION
4 Albert Embankment, London SE1 7SR
www.imo.org

Second edition 2010

Printed by Polestar Wheatons (UK) Ltd, Exeter, EX2 8RP

ISBN: 978-92-801-1507-9

IMO PUBLICATION
Sales number: ID982E

This publication has been prepared from official documents of IMO, and every effort has been made to eliminate errors and reproduce the original text(s) faithfully. Readers should be aware that, in case of inconsistency, the official IMO text will prevail.

075712

Contents

Foreword 1

Resolution MSC.48(66) 5

Preamble 7

Chapter I General 7

1.1 Definitions 7

1.2 General requirements for life-saving appliances 8

Chapter II Personal life-saving appliances 9

2.1 Lifebuoys 9

2.2 Lifejackets 11

2.3 Immersion suits 15

2.4 Anti-exposure suits 17

2.5 Thermal protective aids 19

Chapter III Visual signals 19

3.1 Rocket parachute flares 19

3.2 Hand flares 20

3.3 Buoyant smoke signals 20

Chapter IV Survival craft 21

4.1 General requirements for liferafts 21

4.2 Inflatable liferafts 28

4.3 Rigid liferafts 33

4.4 General requirements for lifeboats 35

4.5 Partially enclosed lifeboats 47

4.6 Totally enclosed lifeboats 48

4.7 Free-fall lifeboats 50

4.8 Lifeboats with a self-contained air support system 53

4.9 Fire-protected lifeboats 53

Chapter V Rescue boats 54

5.1 Rescue boats 54

Chapter VI Launching and embarkation appliances 60
6.1 Launching and embarkation appliances 60
6.2 Marine evacuation systems 66

Chapter VII Other life-saving appliances 69
7.1 Line-throwing appliances 69
7.2 General alarm and public address system 70

Testing and Evaluation of Life-Saving Appliances

I Revised recommendation on testing of life-saving appliances, as amended 75
Resolution MSC.81(70) 77
Introduction 79

Part 1 *Prototype tests for life-saving appliances* 80
1 **Lifebuoys** 80
1.1 Lifebuoys specification 80
1.2 Temperature cycling test 80
1.3 Drop test 81
1.4 Test for oil resistance 81
1.5 Fire test 81
1.6 Flotation test 81
1.7 Strength test 81
1.8 Test for operation with a light and smoke signal 82
1.9 Lifebuoy self-activating smoke signal tests 82

2 **Lifejackets** 83
2.1 Temperature cycling test 83
2.2 Buoyancy test 83
2.3 Fire test 83
2.4 Tests of components other than buoyancy materials 83
2.5 Strength tests 83
2.6 Tests for lifejacket buoyancy material 86
2.7 Donning test 87
2.8 Water performance tests 89
2.9 Infant and children's lifejacket tests 93
2.10 Tests for inflatable lifejackets 95

3	**Immersion suits, anti-exposure suits and thermal protective aids**	101
3.1	Tests common to non-insulated and insulated immersion suits and anti-exposure suits	101
3.2	Thermal protective tests	105
3.3	Thermal protective aids for survival craft	107
4	**Pyrotechnics – rocket parachute flares, hand flares and buoyant smoke signals**	108
4.1	General	108
4.2	Temperature tests	108
4.3	Water and corrosion resistance test	109
4.4	Handling safety test	109
4.5	Safety inspection	109
4.6	Rocket parachute flares test	110
4.7	Hand flares test	110
4.8	Buoyant smoke signals test	111
5	**Liferafts – rigid and inflatable**	112
5.1	Drop test	112
5.2	Jump test	113
5.3	Weight test	113
5.4	Towing test	114
5.5	Mooring out tests	114
5.6	Liferaft painter system test	114
5.7	Loading and seating test	114
5.8	Boarding and closing arrangement test	115
5.9	Stability test	115
5.10	Manoeuvrability test	116
5.11	Swamp test	116
5.12	Canopy closure test	116
5.13	Buoyancy of float-free liferafts	116
5.14	Detailed inspection	117
5.15	Weak link test	117
5.16	Davit-launched liferafts – strength test of lifting components	117
5.17	Additional tests applicable to inflatable liferafts only	118
5.18	Additional tests applicable to automatically self-righting liferafts only	132

5.19	Submergence test for automatically self-righting and canopied reversible liferafts	133
5.20	Wind velocity tests	133
5.21	Test for self-draining of floors of canopied reversible liferafts and automatically self-righting liferafts	134
5.22	Liferaft light tests	134
6	**Lifeboats**	134
6.1	Definitions and general conditions	134
6.2	Lifeboat material tests	134
6.3	Lifeboat overload test	135
6.4	Davit-launched lifeboat impact and drop test	137
6.5	Free-fall lifeboat free-fall test	139
6.6	Lifeboat seating strength test	140
6.7	Lifeboat seating space test	140
6.8	Lifeboat freeboard and stability tests	141
6.9	Release mechanism test	142
6.10	Lifeboat operational test	143
6.11	Lifeboat towing and painter release test	144
6.12	Lifeboat light tests	145
6.13	Canopy erection test	145
6.14	Additional tests for totally enclosed lifeboats	145
6.15	Air supply test for lifeboats with a self-contained air support system	147
6.16	Additional tests for fire-protected lifeboats	148
6.17	Measuring and evaluating acceleration forces	149
7	**Rescue boats and fast rescue boats**	153
7.1	Rigid rescue boats	153
7.2	Inflated rescue boats	155
7.3	Rigid/inflated rescue boats	159
7.4	Rigid fast rescue boats	159
7.5	Inflated fast rescue boats	160
7.6	Rigid/inflated fast rescue boats	160
7.7	Outboard motors for rescue boats	160
8	**Launching and embarkation appliances**	162
8.1	Testing of davits and launching appliances	162
8.2	Davit-launched liferaft automatic release hook test	164

9	**Line-throwing appliances**	167
9.1	Test for pyrotechnics	167
9.2	Function test	168
9.3	Line tensile test	168
9.4	Visual examination	168
9.5	Temperature test	168
10	**Position-indicating lights for life-saving appliances**	168
10.1	Survival craft and rescue boats light tests	168
10.2	Lifebuoy self-igniting light tests	169
10.3	Lifejacket light tests	170
10.4	Common tests for all position-indicating lights	171
11	**Hydrostatic release units**	177
11.1	Visual and dimensional examination	177
11.2	Technical tests	177
11.3	Performance test	179
12	**Marine evacuation systems**	180
12.1	Materials	180
12.2	Marine evacuation system container	180
12.3	Marine evacuation passage	180
12.4	Marine evacuation platform, if fitted	182
12.5	Associated inflatable liferafts	183
12.6	Performance	183
13	**Searchlights for lifeboats and rescue boats**	185
13.1	Visual examination	185
13.2	Durability and resistance to environmental conditions	185
13.3	Operational controls	186
13.4	Light tests	187
Part 2	*Production and installation tests*	188
1	**General**	188
2	**Individual buoyancy equipment**	188
2.1	Lifejackets	188
2.2	Immersion and anti-exposure suits	189
3	**Portable buoyancy equipment**	189
3.1	Lifebuoys	189

4 **Pyrotechnics** . . . 189
5 **Survival craft** . . . 189
5.1 Liferaft operational inflation test . . . 189
5.2 Davit-launched liferaft and inflated rescue boat test . . . 191
5.3 Lifeboat and rescue boat test . . . 191
5.4 Launch test . . . 192
6 **Launching and stowage arrangements** . . . 192
6.1 Launching appliances using falls and winches . . . 192
6.2 Installation tests of liferaft launching appliances . . . 194
7 **Marine evacuation systems** . . . 196
7.1 Installation tests . . . 196

Annex 1 – Adult reference test device (RTD) design and construction . . . 197
1 General . . . 197
2 Materials . . . 197
3 Construction . . . 198

Annex 2 – Child reference test device (RTD) design and construction . . . 219
1 General . . . 219
2 Materials . . . 219
3 Construction . . . 220

Annex 3 – Infant reference test device (RTD) design and construction . . . 237
1 General . . . 237
2 Materials . . . 237
3 Construction . . . 238

II **Code of practice for the evaluation, testing and acceptance of prototype novel life-saving appliances and arrangements** . . . 255
Resolution A.520(13) . . . 257
Preamble . . . 259
1 General provisions . . . 260
2 General criteria . . . 261
3 Appliance criteria and testing of prototypes . . . 266

Foreword

This publication contains the three most important IMO instruments dealing with life-saving appliances, namely the International Life-Saving Appliance (LSA) Code, the Revised Recommendation on Testing of Life-Saving Appliances and the Code of Practice for the Evaluation, Testing and Acceptance of Prototype Novel Life-Saving Appliances.

The International Life-Saving Appliance (LSA) Code was adopted by IMO's Maritime Safety Committee (MSC) at its 66th session (June 1996) by resolution MSC.48(66). It provides international requirements for the life-saving appliances required by chapter III of the 1974 SOLAS Convention, including personal life-saving appliances like lifebuoys, lifejackets, immersion suits, anti-exposure suits and thermal protective aids; visual aids, such as parachute flares, hand flares and buoyant smoke signals; survival craft, such as liferafts and lifeboats; rescue boats; launching and embarkation appliances and marine evacuation systems line throwing appliances; and general alarm and public address systems.

The Code was made mandatory by resolution MSC.47(66) under SOLAS regulation III/3.10, whereby regulation III/34 determines that all life-saving appliances and arrangements shall comply with its requirements. The Code entered into force on 1 July 1998 and has been amended in accordance with SOLAS Article VIII as follows:

.1 by the May 2006 amendments, which were adopted by resolution MSC.207(81) and entered into force on 1 July 2010;

.2 by the December 2006 amendments, which were adopted by resolution MSC.218(82) and entered into force on 1 July 2008; and

.3 by the 2008 amendments, which were adopted by resolution MSC.272(85) and entered into force on 1 July 2010.

The consolidated text of the LSA Code in the present publication incorporates the above three sets of amendments.

Recommendations on the testing of life-saving appliances were first adopted by the IMO Assembly in 1991, by resolution A.689(17). In 1998, the MSC, recognizing the need to introduce more precise requirements for the testing of life-saving appliances and recalling that it had amended the recommendations on several occasions since their adoption, adopted the Revised Recommendation on Testing of Life-Saving Appliances (resolution MSC.81(70)), effectively replacing resolution A.689(17). Since then, the Revised Recommendations have again been amended several times, in the main corresponding to the associated LSA Code amendments described above, and the present publication contains the consolidated text including the amendments adopted by MSC 80 (resolution MSC.200(80)), MSC 82 (resolution MSC.226(82)) and MSC 85 (resolution MSC.274(85)).

The Code of practice for the evaluation, testing and acceptance of prototype novel life-saving appliances and arrangements, adopted by the Assembly in 1983 by resolution A.520(13), is intended to cater for prototype novel life-saving appliances and arrangements which may be developed, and do not fully meet the requirements of chapter III of the 1974 SOLAS Convention, but provide the same or higher safety standards.

International Life-Saving Appliance (LSA) Code

Resolution MSC.48(66)

(adopted on 4 June 1996)

THE MARITIME SAFETY COMMITTEE,

RECALLING Article 28(b) of the Convention on the International Maritime Organization concerning the functions of the Committee,

RECOGNIZING the need to provide international standards for life-saving appliances required by chapter III of the International Convention for the Safety of Life at Sea (SOLAS), 1974, as amended,

NOTING resolution MSC.47(66) by which it adopted, *inter alia,* amendments to chapter III of the SOLAS Convention to make the provisions of the International Life-Saving Appliance (LSA) Code mandatory under that Convention on or after 1 July 1998,

HAVING CONSIDERED, at its sixty-sixth session, the text of the proposed LSA Code,

1. ADOPTS the International Life-Saving Appliance (LSA) Code, the text of which is set out in the annex to the present resolution;

2. NOTES that under the amendments to chapter III of the 1974 SOLAS Convention, amendments to the LSA Code shall be adopted, brought into force and shall take effect in accordance with the provisions of article VIII of that Convention concerning the amendments procedure applicable to the annex to the Convention other than chapter I;

3. REQUESTS the Secretary-General to transmit certified copies of the present resolution and the text of the LSA Code contained in the annex to all Contracting Governments to the Convention;

4. FURTHER REQUESTS the Secretary-General to transmit copies of this resolution and its annex to Members of the Organization, which are not Contracting Governments to the Convention.

Preamble

1 The purpose of this Code is to provide international standards for life-saving appliances required by chapter III of the International Convention for the Safety of Life at Sea (SOLAS), 1974.

2 On and after 1 July 1998, the requirements of this Code will be mandatory under the International Convention for the Safety of Life at Sea (SOLAS), 1974, as amended. Any future amendment to the Code will be adopted and brought into force in accordance with the procedure laid down in article VIII of that Convention.

Chapter I
General

1.1 Definitions

1.1.1 *Convention* means the International Convention for the Safety of Life at Sea, 1974, as amended.

1.1.2 *Effective clearing of the ship* is the ability of the free-fall lifeboat to move away from the ship after free-fall launching without using its engine.

1.1.3 *Free-fall acceleration* is the rate of change of velocity experienced by the occupants during launching of a free-fall lifeboat.

1.1.4 *Free-fall certification height* is the greatest launching height for which the lifeboat is to be approved, measured from the still water surface to the lowest point on the lifeboat when the lifeboat is in the launch configuration.

1.1.5 *Launching ramp angle* is the angle between the horizontal and the launch rail of the lifeboat in its launching position with the ship on even keel.

1.1.6 *Launching ramp length* is the distance between the stern of the lifeboat and the lower end of the launching ramp.

1.1.7 *Regulation* means a regulation contained in the annex to the Convention.

1.1.8 *Retro-reflective material* is a material which reflects in the opposite direction a beam of light directed on it.

1.1.9 *Water-entry angle* is the angle between the horizontal and the launch rail of the lifeboat when it first enters the water.

1.1.10 The terms used in this Code have the same meaning as those defined in regulation III/3.

1.2 General requirements for life-saving appliances

1.2.1 Paragraph 1.2.2.7 applies to life-saving appliances on all ships.

1.2.2 Unless expressly provided otherwise or unless, in the opinion of the Administration having regard to the particular voyages on which the ship is constantly engaged, other requirements are appropriate, all life-saving appliances prescribed in this part shall:

.1 be constructed with proper workmanship and materials;

.2 not be damaged in stowage throughout the air temperature range −30°C to +65°C and, in the case of personal life-saving appliances, unless otherwise specified, remain operational throughout the air temperature range of −15°C to +40°C;

.3 if they are likely to be immersed in seawater during their use, operate throughout the seawater temperature range −1°C to +30°C;

.4 where applicable, be rot-proof, corrosion-resistant, and not be unduly affected by seawater, oil or fungal attack;

.5 where exposed to sunlight, be resistant to deterioration;

.6 be of an international or vivid reddish orange, or at a comparably highly visible colour on all parts where this will assist detection at sea;

.7 be fitted with retro-reflective material where it will assist in detection and in accordance with the recommendations of the Organization;*

* Refer to the Recommendation on the use and fitting of retro-reflective materials on life-saving appliances adopted by the Organization by resolution A.658(16), as it may be amended.

.8 if they are to be used in a seaway, be capable of satisfactory operation in that environment;

.9 be clearly marked with approval information, including the Administration which approved it and any operational restrictions; and

.10 where applicable, be provided with electrical short-circuit protection to prevent damage or injury.

1.2.3 The Administration shall determine the period of acceptability of life-saving appliances which are subject to deterioration with age. Such life-saving appliances shall be marked with a means for determining their age or the date by which they must be replaced. Permanent marking with a date of expiry is the preferred method of establishing the period of acceptability. Batteries not marked with an expiration date may be used if they are replaced annually, or in the case of a secondary battery (accumulator), if the condition of the electrolyte can be readily checked. In the case of pyrotechnic life-saving appliances, the date of expiry shall be indelibly marked on the product by the manufacturer.

Chapter II
Personal life-saving appliances

2.1 Lifebuoys

2.1.1 *Lifebuoy specification*

Every lifebuoy shall:

.1 have an outer diameter of not more than 800 mm and an inner diameter of not less than 400 mm;

.2 be constructed of inherently buoyant material; it shall not depend upon rushes, cork shavings or granulated cork, any other loose granulated material or any air compartment which depends on inflation for buoyancy;

.3 be capable of supporting not less than 14.5 kg of iron in fresh water for a period of 24 h;

.4 have a mass of not less than 2.5 kg;

.5 not sustain burning or continue melting after being totally enveloped in a fire for a period of 2 s;

.6 be constructed to withstand a drop into the water from the height at which it is stowed above the waterline in the lightest seagoing condition or 30 m, whichever is the greater, without impairing either its operating capability or that of its attached components;

.7 if it is intended to operate the quick-release arrangement provided for the self-activated smoke signals and self-igniting lights, have a mass of not less than 4 kg; and

.8 be fitted with a grabline not less than 9.5 mm in diameter and not less than four times the outside diameter of the body of the buoy in length. The grabline shall be secured at four equidistant points around the circumference of the buoy to form four equal loops.

2.1.2 *Lifebuoy self-igniting lights*

Self-igniting lights required by regulation III/7.1.3 shall:

.1 be such that they cannot be extinguished by water;

.2 be of white colour and capable of either burning continuously with a luminous intensity of not less than 2 cd in all directions of the upper hemisphere or flashing (discharge flashing) at a rate of not less than 50 flashes and not more than 70 flashes per minute with at least the corresponding effective luminous intensity;

.3 be provided with a source of energy capable of meeting the requirement of paragraph 2.1.2.2 for a period of at least 2 h; and

.4 be capable of withstanding the drop test required by paragraph 2.1.1.6.

2.1.3 *Lifebuoy self-activating smoke signals*

Self-activating smoke signals required by regulation III/7.1.3 shall:

.1 emit smoke of a highly visible colour at a uniform rate for a period of at least 15 min when floating in calm water;

.2 not ignite explosively or emit any flame during the entire smoke emission time of the signal;

.3 not be swamped in a seaway;

.4 continue to emit smoke when fully submerged in water for a period of at least 10 s;

.5 be capable of withstanding the drop test required by paragraph 2.1.1.6; and

.6 be provided with a quick-release arrangement that will automatically release and activate the signal and associated self-igniting light connected to a lifebuoy having a mass of not more than 4 kg.

2.1.4 *Buoyant lifelines*

Buoyant lifelines required by regulation III/7.1.2 shall:

.1 be non-kinking;

.2 have a diameter of not less than 8 mm; and

.3 have a breaking strength of not less than 5 kN.

2.2 Lifejackets

2.2.1 *General requirements for lifejackets*

2.2.1.1 A lifejacket shall not sustain burning or continue melting after being totally enveloped in a fire for a period of 2 s.

2.2.1.2 Lifejackets shall be provided in three sizes in accordance with table 2.1. If a lifejacket fully complies with the requirements of two adjacent size ranges, it may be marked with both size ranges, but the specified ranges shall not be divided. Lifejackets shall be marked by either weight or height, or by both weight and height, according to table 2.1.

Table 2.1 – *Lifejacket sizing criteria*

Lifejacket marking	Infant	Child	Adult
User's size:			
Weight (kg)	Less than 15	15 or more but less than 43	43 or more
Height (cm)	Less than 100	100 or more but less than 155	155 or more

2.2.1.3 If an adult lifejacket is not designed to fit persons weighing up to 140 kg and with a chest girth of up to 1,750 mm, suitable accessories shall be available to allow it to be secured to such persons.

2.2.1.4 The in-water performance of a lifejacket shall be evaluated by comparison to the performance of a suitable size standard reference lifejacket, i.e., reference test device (RTD) complying with the recommendations of the Organization.*

2.2.1.5 An adult lifejacket shall be so constructed that:

.1 at least 75% of persons who are completely unfamiliar with the lifejacket can correctly don it within a period of 1 min without assistance, guidance or prior demonstration;

.2 after demonstration, all persons can correctly don it within a period of 1 min without assistance;

.3 it is clearly capable of being worn in only one way or inside-out and, if donned incorrectly, it is not injurious to the wearer;

.4 the method of securing the lifejacket to the wearer has quick and positive means of closure that do not require tying of knots;

.5 it is comfortable to wear; and

.6 it allows the wearer to jump into the water from a height of at least 4.5 m while holding on to the lifejacket, and from a height of at least 1 m with arms held overhead, without injury and without dislodging or damaging the lifejacket or its attachments.

2.2.1.6 When tested according to the recommendations of the Organization on at least 12 persons, adult lifejackets shall have sufficient buoyancy and stability in calm fresh water to:

.1 lift the mouth of exhausted or unconscious persons by an average height of not less than the average provided by the adult RTD;

.2 turn the body of unconscious, face-down persons in the water to a position where the mouth is clear of the water in an average time not exceeding that of the RTD, with the number of persons not turned by the lifejacket no greater than that of the RTD;

* Refer to the Revised Recommendation on testing of life-saving appliances (resolution MSC.81(70)), as amended.

.3 incline the body backwards from the vertical position for an average torso angle of not less than that of the RTD minus 5°;

.4 lift the head above horizontal for an average faceplane angle of not less than that of the RTD minus 5°; and

.5 return the wearer to a stable face-up position after being destabilized when floating in the flexed foetal position.*

2.2.1.7 An adult lifejacket shall allow the person wearing it to swim a short distance and to board a survival craft.

2.2.1.8 An infant or child lifejacket shall perform the same as an adult lifejacket except as follows:

.1 donning assistance is permitted for small children and infants;

.2 the appropriate child or infant RTD shall be used in place of the adult RTD; and

.3 assistance may be given to board a survival craft, but wearer mobility shall not be reduced to any greater extent than by the appropriate size RTD.

2.2.1.9 With the exception of freeboard and self-righting performance, the requirements for infant lifejackets may be relaxed, if necessary, in order to:

.1 facilitate the rescue of the infant by a caretaker;

.2 allow the infant to be fastened to a caretaker and contribute to keeping the infant close to the caretaker;

.3 keep the infant dry, with free respiratory passages;

.4 protect the infant against bumps and jolts during evacuation; and

.5 allow a caretaker to monitor and control heat loss by the infant.

2.2.1.10 In addition to the markings required by paragraph 1.2.2.9, an infant or child lifejacket shall be marked with:

.1 the size range in accordance with paragraph 2.2.1.2; and

---

* Refer to the illustration on page 11 of the *IMO Pocket Guide to Cold Water Survival* and to the Revised Recommendation on testing of life-saving appliances (resolution MSC.81(70)), as amended.

.2 an "infant" or "child" symbol as shown in the "infant's lifejacket" or "child's lifejacket" symbol adopted by the Organization.*

2.2.1.11 A lifejacket shall have buoyancy which is not reduced by more than 5% after 24 h submersion in fresh water.

2.2.1.12 The buoyancy of a lifejacket shall not depend on the use of loose granulated materials.

2.2.1.13 Each lifejacket shall be provided with means of securing a lifejacket light as specified in paragraph 2.2.3 such that it shall be capable of complying with paragraphs 2.2.1.5.6 and 2.2.3.1.3.

2.2.1.14 Each lifejacket shall be fitted with a whistle firmly secured by a lanyard.

2.2.1.15 Lifejacket lights and whistles shall be selected and secured to the lifejacket in such a way that their performance in combination is not degraded.

2.2.1.16 A lifejacket shall be provided with a releasable buoyant line or other means to secure it to a lifejacket worn by another person in the water.

2.2.1.17 A lifejacket shall be provided with a suitable means to allow a rescuer to lift the wearer from the water into a survival craft or rescue boat.

2.2.2 *Inflatable lifejackets*

A lifejacket which depends on inflation for buoyancy shall have not less than two separate compartments shall comply with the requirements of paragraph 2.2.1 and shall:

.1 inflate automatically upon immersion, be provided with a device to permit inflation by a single manual motion and be capable of having each chamber inflated by mouth;

.2 in the event of loss of buoyancy in any one compartment be capable of complying with the requirements of paragraphs 2.2.1.5, 2.2.1.6 and 2.2.1.7; and

.3 comply with the requirements of paragraph 2.2.1.11 after inflation by means of the automatic mechanism.

* Refer to Symbols related to life-saving appliances and arrangements, adopted by the Organization by resolution A.760(18), as amended.

2.2.3 *Lifejacket lights*

2.2.3.1 Each lifejacket light shall:

- **.1** have a luminous intensity of not l
tions of the upper hemisphere;
- **.2** have a source of energy capabl
intensity of 0.75 cd for a period of
- **.3** be visible over as great a segment o
is practicable when attached to a lif
- **.4** be of white colour.

2.2.3.2 If the light referred to in paragraph 2.2.3.1 is a flashing light, it shall, in addition:

- **.1** be provided with a manually operated switch; and
- **.2** flash at a rate of not less than 50 flashes and not more than 70 flashes per minute with an effective luminous intensity of at least 0.75 cd.

2.3 Immersion suits

2.3.1 *General requirements for immersion suits*

2.3.1.1 An immersion suit shall be constructed with waterproof materials such that:

- **.1** it can be unpacked and donned without assistance within 2 min, taking into account donning of any associated clothing, donning of a lifejacket if the immersion suit is to be worn in conjunction with a lifejacket to meet the requirements of paragraph 2.3.1.2, and inflation of orally inflatable chambers if fitted;*
- **.2** it will not sustain burning or continue melting after being totally enveloped in a fire for a period of 2 s;
- **.3** it will cover the whole body with the exception of the face, except that covering for the hands may be provided by separate gloves which shall be permanently attached to the suit;
- **.4** it is provided with arrangements to minimize or reduce free air in the legs of the suit; and

* Refer to paragraph 3.1.3 of the Revised Recommendation on testing of life-saving appliances, adopted by the MSC by resolution MSC.81(70), as amended.

following a jump from a height of not less than 4.5 m into the water there is no undue ingress of water into the suit.

.1.2 An immersion suit on its own, or worn in conjunction with a lifejacket if necessary, shall have sufficient buoyancy and stability in calm fresh water to:

.1 lift the mouth of an exhausted or unconscious person clear of the water by not less than 120 mm; and

.2 allow the wearer to turn from a face-down to a face-up position in not more than 5 s.

2.3.1.3 An immersion suit shall permit the person wearing it, and also wearing a lifejacket if the immersion suit is to be worn in conjunction with a lifejacket, to:

.1 climb up and down a vertical ladder at least 5 m in length;

.2 perform normal duties associated with abandonment;

.3 jump from a height of not less than 4.5 m into the water without damaging or dislodging the immersion suit or its attachments, or being injured; and

.4 swim a short distance through the water and board a survival craft.

2.3.1.4 An immersion suit which has buoyancy and is designed to be worn without a lifejacket shall be fitted with a light complying with the requirements of paragraph 2.2.3 and the whistle prescribed by paragraph 2.2.1.14.

2.3.1.5 An immersion suit which has buoyancy and is designed to be worn without a lifejacket shall be provided with a releasable buoyant line or other means to secure it to a suit worn by another person in the water.

2.3.1.6 An immersion suit which has buoyancy and is designed to be worn without a lifejacket shall be provided with a suitable means to allow a rescuer to lift the wearer from the water into a survival craft or rescue boat.

2.3.1.7 If an immersion suit is to be worn in conjunction with a lifejacket, the lifejacket shall be worn over the immersion suit. Persons wearing such an immersion suit shall be able to don a lifejacket without assistance. The immersion suit shall be marked to indicate that it must be worn in conjunction with a compatible lifejacket.

2.3.1.8 An immersion suit shall have buoyancy which is not reduced by more than 5% after 24 h submersion in fresh water and does not depend on the use of loose granulated materials.

2.3.2 *Thermal performance requirements for immersion suits*

2.3.2.1 An immersion suit made of material which has no inherent insulation shall be:

.1 marked with instructions that it must be worn in conjunction with warm clothing; and

.2 so constructed that, when worn in conjunction with warm clothing, and with a lifejacket if the immersion suit is to be worn with a lifejacket, the immersion suit continues to provide sufficient thermal protection, following one jump by the wearer into the water from a height of 4.5 m, to ensure that when it is worn for a period of 1 h in calm circulating water at a temperature of 5°C, the wearer's body core temperature does not fall more than 2°C.

2.3.2.2 An immersion suit made of material with inherent insulation, when worn either on its own or with a lifejacket, if the immersion suit is to be worn in conjunction with a lifejacket, shall provide the wearer with sufficient thermal insulation, following one jump into the water from a height of 4.5 m, to ensure that the wearer's body core temperature does not fall more than 2°C after a period of 6 h immersion in calm circulating water at a temperature of between 0°C and 2°C.

2.4 Anti-exposure suits

2.4.1 *General requirements for anti-exposure suits*

2.4.1.1 An anti-exposure suit shall be constructed with waterproof materials such that it:

.1 provides inherent buoyancy of at least 70 N;

.2 is made of material which reduces the risk of heat stress during rescue and evacuation operations;

.3 covers the whole body except, where the Administration so permits, the feet; covering for the hands and head may be provided by separate gloves and a hood, both of which shall be permanently attached to the suit;

.4 can be unpacked and donned without assistance within 2 min;

.5 does not sustain burning or continue melting after being totally enveloped in a fire for a period of 2 s;

.6 is equipped with a pocket for a portable VHF telephone; and

.7 has a lateral field of vision of at least 120°.

2.4.1.2 An anti-exposure suit shall permit the person wearing it to:

.1 climb up and down a vertical ladder of at least 5 m in length;

.2 jump from a height of not less than 4.5 m into the water with feet first, without damaging or dislodging the suit or its attachments, or being injured;

.3 swim through the water at least 25 m and board a survival craft;

.4 don a lifejacket without assistance; and

.5 perform all duties associated with abandonment, assist others and operate a rescue boat.

2.4.1.3 An anti-exposure suit shall be fitted with a light complying with the requirements of paragraph 2.2.3 such that it shall be capable of complying with paragraphs 2.2.3.1.3 and 2.4.1.2.2, and the whistle prescribed by paragraph 2.2.1.14.

2.4.2 ***Thermal performance requirements for anti-exposure suits***

2.4.2.1 An anti-exposure suit shall:

.1 if made of material which has no inherent insulation, be marked with instructions that it must be worn in conjunction with warm clothing; and

.2 be so constructed that, when worn as marked and following one jump into the water that totally submerges the wearer, the suit continues to provide sufficient thermal protection to ensure that, when it is worn in calm circulating water at a temperature of 5°C, the wearer's body core temperature does not fall at a rate of more than 1.5°C per hour after the first 0.5 h.

2.4.3 *Stability requirements*

A person in fresh water wearing an anti-exposure suit complying with the requirements of this section shall be able to turn from a face-down to a face-up position in not more than 5 s and shall be stable face-up. The suit shall have no tendency to turn the wearer face-down in moderate sea conditions.

2.5 Thermal protective aids

2.5.1 A thermal protective aid shall be made of waterproof material having a thermal conductance of not more than 7,800 W/(m^2 K) and shall be so constructed that, when used to enclose a person, it shall reduce both the convective and evaporative heat loss from the wearer's body.

2.5.2 The thermal protective aid shall:

- **.1** cover the whole body of persons of all sizes wearing a lifejacket with the exception of the face. Hands shall also be covered unless permanently attached gloves are provided;
- **.2** be capable of being unpacked and easily donned without assistance in a survival craft or rescue boat; and
- **.3** permit the wearer to remove it in the water in not more than 2 min, if it impairs ability to swim.

2.5.3 The thermal protective aid shall function properly throughout an air temperature range –30°C to +20°C.

Chapter III
Visual signals

3.1 Rocket parachute flares

3.1.1 The rocket parachute flare shall:

- **.1** be contained in a water-resistant casing;
- **.2** have brief instructions or diagrams clearly illustrating the use of the rocket parachute flare printed on its casing;
- **.3** have integral means of ignition; and
- **.4** be so designed as not to cause discomfort to the person holding the casing when used in accordance with the manufacturer's operating instructions.

3.1.2 The rocket shall, when fired vertically, reach an altitude of not less than 300 m. At or near the top of its trajectory, the rocket shall eject a parachute flare, which shall:

.1 burn with a bright red colour;

.2 burn uniformly with an average luminous intensity of not less than 30,000 cd;

.3 have a burning period of not less than 40 s;

.4 have a rate of descent of not more than 5 m/s; and

.5 not damage its parachute or attachments while burning.

3.2 Hand flares

3.2.1 The hand flare shall:

.1 be contained in a water-resistant casing;

.2 have brief instructions or diagrams clearly illustrating the use of the hand flare printed on its casing;

.3 have a self-contained means of ignition; and

.4 be so designed as not to cause discomfort to the person holding the casing and not endanger the survival craft by burning or glowing residues when used in accordance with the manufacturer's operating instructions.

3.2.2 The hand flare shall:

.1 burn with a bright red colour;

.2 burn uniformly with an average luminous intensity of not less than 15,000 cd;

.3 have a burning period of not less than 1 min; and

.4 continue to burn after having been immersed for a period of 10 s under 100 mm of water.

3.3 Buoyant smoke signals

3.3.1 The buoyant smoke signal shall:

.1 be contained in a water-resistant casing;

.2 not ignite explosively when used in accordance with the manufacturer's operating instructions; and

.3 have brief instructions or diagrams clearly illustrating the use of the buoyant smoke signal printed on its casing.

3.3.2 The buoyant smoke signal shall:

.1 emit smoke of a highly visible colour at a uniform rate for a period of not less than 3 min when floating in calm water;

.2 not emit any flame during the entire smoke emission time;

.3 not be swamped in a seaway; and

.4 continue to emit smoke when submerged in water for a period of 10 s under 100 mm of water.

Chapter IV
Survival craft

4.1 General requirements for liferafts

4.1.1 *Construction of liferafts*

4.1.1.1 Every liferaft shall be so constructed as to be capable of withstanding exposure for 30 days afloat in all sea conditions.

4.1.1.2 The liferaft shall be so constructed that when it is dropped into the water from a height of 18 m, the liferaft and its equipment will operate satisfactorily. If the liferaft is to be stowed at a height of more than 18 m above the waterline in the lightest seagoing condition, it shall be of a type which has been satisfactorily drop-tested from at least that height.

4.1.1.3 The floating liferaft shall be capable of withstanding repeated jumps onto it from a height of at least 4.5 m above its floor both with and without the canopy erected.

4.1.1.4 The liferaft and its fittings shall be so constructed as to enable it to be towed at a speed of 3 knots in calm water when loaded with its full complement of persons and equipment and with one of its sea-anchors streamed.

4.1.1.5 The liferaft shall have a canopy to protect the occupants from exposure which is automatically set in place when the liferaft is launched and waterborne. The canopy shall comply with the following:

- **.1** it shall provide insulation against heat and cold by means of either two layers of material separated by an air gap or other equally efficient means. Means shall be provided to prevent accumulation of water in the air gap;
- **.2** its interior shall be of a colour that does not cause discomfort to the occupants;
- **.3** each entrance shall be clearly indicated and be provided with efficient adjustable closing arrangements which can be easily and quickly opened by persons clothed in immersion suits from inside and outside, and closed from inside the liferaft so as to permit ventilation but exclude seawater, wind and cold. Liferafts accommodating more than eight persons shall have at least two diametrically opposite entrances;
- **.4** it shall admit sufficient air for the occupants at all times, even with the entrances closed;
- **.5** it shall be provided with at least one viewing port;
- **.6** it shall be provided with means for collecting rainwater;
- **.7** it shall be provided with means to mount a survival craft radar transponder at a height of at least 1 m above the sea; and
- **.8** it shall have sufficient headroom for sitting occupants under all parts of the canopy.

4.1.2 *Minimum carrying capacity and mass of liferafts*

4.1.2.1 No liferaft shall be approved which has a carrying capacity of less than six persons calculated in accordance with the requirements of paragraph 4.2.3 or 4.3.3, as appropriate.

4.1.2.2 Unless the liferaft is to be launched by an approved launching appliance complying with the requirements of section 6.1 or is not intended for easy side-to-side transfer, the total mass of the liferaft, its container and its equipment shall not be more than 185 kg.

4.1.3 *Liferaft fittings*

4.1.3.1 Lifelines shall be securely becketed around the inside and outside of the liferaft.

4.1.3.2 The liferaft shall be fitted with an efficient painter of length equal to not less than 10 m plus the distance from the stowed position to the waterline in the lightest seagoing condition or 15 m, whichever is the greater. The breaking strength of the painter system, including its means of attachment to the liferaft, except the weak link required by paragraph 4.1.6, shall be not less than 15 kN for liferafts permitted to accommodate more than 25 persons, not less than 10 kN for liferafts permitted to accommodate 9 to 25 persons and not less than 7.5 kN for any other liferaft.

4.1.3.3 A manually controlled exterior light shall be fitted to the uppermost portion of the liferaft canopy or structure. The light shall be white and be capable of operating continuously for at least 12 h with a luminous intensity of not less than 4.3 cd in all directions of the upper hemisphere. However, if the light is a flashing light it shall flash at a rate of not less than 50 flashes and not more than 70 flashes per minute for the 12 h operating period with an equivalent effective luminous intensity. The lamp shall light automatically when the canopy is erected. Batteries shall be of a type that does not deteriorate due to dampness or humidity in the stowed liferaft.

4.1.3.4 A manually controlled interior light shall be fitted inside the liferaft capable of continuous operation for a period of at least 12 h. It shall light automatically when the canopy is erected and shall produce an arithmetic mean luminous intensity of not less than 0.5 cd when measured over the entire upper hemisphere to permit reading of survival and equipment instructions. Batteries shall be of a type that does not deteriorate due to damp or humidity in the stowed liferaft.

4.1.4 *Davit-launched liferafts*

4.1.4.1 In addition to the above requirements, a liferaft for use with an approved launching appliance shall:

- **.1** when the liferaft is loaded with its full complement of persons and equipment, be capable of withstanding a lateral impact against the ship's side at an impact velocity of not less than 3.5 m/s and also a drop into the water from a height of not less than 3 m without damage that will affect its function;
- **.2** be provided with means for bringing the liferaft alongside the embarkation deck and holding it securely during embarkation.

4.1.4.2 Every passenger ship davit-launched liferaft shall be so arranged that it can be rapidly boarded by its full complement of persons.

4.1.4.3 Every cargo ship davit-launched liferaft shall be so arranged that it can be boarded by its full complement of persons in not more than 3 min from the time the instruction to board is given.

4.1.5 *Equipment*

4.1.5.1 The normal equipment of every liferaft shall consist of:

.1 one buoyant rescue quoit, attached to not less than 30 m of buoyant line;

.2 one knife of the non-folding type having a buoyant handle and lanyard attached and stowed in a pocket on the exterior of the canopy near the point at which the painter is attached to the liferaft. In addition, a liferaft which is permitted to accommodate 13 persons or more shall be provided with a second knife which need not be of the non-folding type;

.3 for a liferaft which is permitted to accommodate not more than 12 persons, one buoyant bailer. For a liferaft which is permitted to accommodate 13 persons or more, two buoyant bailers;

.4 two sponges;

.5 two sea-anchors each with a shock-resistant hawser and tripping line if fitted, one being spare and the other per-manently attached to the liferaft in such a way that when the liferaft inflates or is waterborne it will cause the liferaft to lie oriented to the wind in the most stable manner. The strength of each sea-anchor and its hawser and tripping line if fitted shall be adequate in all sea conditions. The sea-anchors shall have means to prevent twisting of the line and shall be of a type which is unlikely to turn inside out between its shroud lines. The sea-anchor permanently attached to davit-launched liferafts and liferafts fitted on passenger ships shall be arranged for manual deployment only. All other liferafts are to have the sea-anchor deployed automatically when the liferaft inflates;

.6 two buoyant paddles;

.7 three tin-openers and a pair of scissors. Safety knives containing special tin-opener blades are satisfactory for this requirement;

.8 one first-aid outfit in a waterproof case capable of being closed tightly after use;

.9 one whistle or equivalent sound signal;

.10 four rocket parachute flares complying with the requirements of section 3.1;

.11 six hand flares complying with the requirements of section 3.2;

.12 two buoyant smoke signals complying with the requirements of section 3.3;

.13 one waterproof electric torch suitable for Morse signalling together with one spare set of batteries and one spare bulb in a waterproof container;

.14 an efficient radar reflector, unless a survival craft radar transponder is stowed in the liferaft;

.15 one daylight signalling mirror with instructions on its use for signalling to ships and aircraft;

.16 one copy of the life-saving signals referred to in regulation V/29 on a waterproof card or in a waterproof container;

.17 one set of fishing tackle;

.18 a food ration consisting of not less than 10,000 kJ (2,400 kcal) for each person the liferaft is permitted to accommodate. These rations shall be palatable, edible throughout the marked life, and packed in a manner which can be readily divided and easily opened, taking into account immersion suit gloved hands.*

The rations shall be packed in permanently sealed metal containers or vacuum packed in a flexible packaging material with a negligible vapour transmission rate (< 0.1 g/m^2 per 24 h at 23°C/85% relative humidity) when tested to a standard

* Note: A typical suitable composition is:
Ration unit: 500–550 g
Energy: Minimum 10,000 kJ
Moisture: Maximum 5%
Salt (NaCl): Maximum 0.2%
Carbohydrates: 60–70% weight = 50–60% energy
Fat: 18–23% weight = 33–43% energy
Protein: 6–10% weight = 5–8% energy

acceptable to the Administration. Flexible packaging materials shall be further protected by outer packaging, if needed, to prevent physical damage to the food ration and other items as result of sharp edges. The packaging shall be clearly marked with date of packing and date of expiry, the production lot number, the content in the package and instructions for use. Food rations complying with the requirements of an international standard acceptable to the Organization* are acceptable in compliance with these requirements;

.19 1.5 ℓ of fresh water for each person the liferaft is permitted to accommodate, of which either 0.5 ℓ per person may be replaced by a de-salting apparatus capable of producing an equal amount of fresh water in two days or 1 ℓ per person may be replaced by a manually powered reverse osmosis desalinator, as described in paragraph 4.4.7.5, capable of producing an equal amount of fresh water in two days. The water shall satisfy suitable international requirements for chemical and microbiological content, and shall be packed in sealed watertight containers that are of corrosion resistant material or are treated to be corrosion resistant. Flexible packaging materials, if used, shall have a negligible vapour transmission rate (< 0.1 g/m^2 per 24 hours at 23°C/85% relative humidity when tested to a standard acceptable to the Administration, except that individually packaged portions within a larger container need not meet this vapour transmission requirement. Each water container shall have a method of spill proof reclosure, except for individually packaged portions of less than 125 mℓ. Each container shall be clearly marked with date of packing and date of expiry, the production lot number, the quantity of water in the container, and instructions for consumption. The containers shall be easy to open, taking into account immersion suit gloved hands. Water for emergency drinking complying with the requirements of an international standard acceptable to the Organization* is acceptable in compliance with these requirements;

.20 one rustproof graduated drinking vessel;

* Refer to the recommendations of the International Organization for Standardization, in particular publication ISO 18813:2006 *Ships and marine technology – Survival equipment for survival craft and rescue boats*.

.21 anti-seasickness medicine sufficient for at least 48 h and one seasickness bag for each person the liferaft is permitted to accommodate;

.22 instructions on how to survive;*

.23 instructions for immediate action; and

.24 thermal protective aids complying with the requirements of section 2.5 sufficient for 10% of the number of persons the liferaft is permitted to accommodate or two, whichever is the greater.

4.1.5.2 The marking required by paragraphs 4.2.6.3.5 and 4.3.6.7 on liferafts equipped in accordance with paragraph 4.1.5.1 shall be "SOLAS A PACK" in block capitals of the Roman alphabet.

4.1.5.3 In the case of passenger ships engaged on short international voyages of such a nature and duration that, in the opinion of the Administration, not all of the items specified in paragraph 4.1.5.1 are necessary, the Administration may allow the liferafts carried on any such ships to be provided with the equipment specified in paragraphs 4.1.5.1.1 to 4.1.5.1.6 inclusive, 4.1.5.1.8, 4.1.5.1.9, 4.1.5.1.13 to 4.1.5.1.16 inclusive and 4.1.5.1.21 to 4.1.5.1.24 inclusive and one half of the equipment specified in paragraphs 4.1.5.1.10 to 4.1.5.1.12 inclusive. The marking required by paragraphs 4.2.6.3.5 and 4.3.6.7 on such liferafts shall be "SOLAS B PACK" in block capitals of the Roman alphabet.

4.1.5.4 Where appropriate the equipment shall be stowed in a container which, if it is not an integral part of, or permanently attached to, the liferaft, shall be stowed and secured inside the liferaft and be capable of floating in water for at least 30 min without damage to its contents.

4.1.6 *Float-free arrangements for liferafts*

4.1.6.1 *Painter system*

The liferaft painter system shall provide a connection between the ship and the liferaft and shall be so arranged as to ensure that the liferaft when released and, in the case of an inflatable liferaft, inflated is not dragged under by the sinking ship.

* Refer to the Instructions for action in survival craft, adopted by the Organization by resolution A.657(16).

4.1.6.2 *Weak link*

If a weak link is used in the float-free arrangement, it shall:

.1 not be broken by the force required to pull the painter from the liferaft container;

.2 if applicable, be of sufficient strength to permit the inflation of the liferaft; and

.3 break under a strain of 2.2 ± 0.4 kN.

4.1.6.3 *Hydrostatic release units*

If a hydrostatic release unit is used in the float-free arrangements, it shall:

.1 be constructed of compatible materials so as to prevent malfunction of the unit. Galvanizing or other forms of metallic coating on parts of the hydrostatic release unit shall not be accepted;

.2 automatically release the liferaft at a depth of not more than 4 m;

.3 have drains to prevent the accumulation of water in the hydro-static chamber when the unit is in its normal position;

.4 be so constructed as to prevent release when seas wash over the unit;

.5 be permanently marked on its exterior with its type and serial number;

.6 be permanently marked, on the unit or identification plate securely attached to the unit, with the date of manufacture, type and serial number and whether the unit is suitable for use with a liferaft with a capacity of more than 25 persons;

.7 be such that each part connected to the painter system has a strength of not less than that required for the painter; and

.8 if disposable, in lieu of the requirement in paragraph 4.1.6.3.6, be marked with a means of determining its date of expiry.

4.2 Inflatable liferafts

4.2.1 Inflatable liferafts shall comply with the requirements of section 4.1 and, in addition, shall comply with the requirements of this section.

4.2.2 *Construction of inflatable liferafts*

4.2.2.1 The main buoyancy chamber shall be divided into not less than two separate compartments, each inflated through a nonreturn inflation valve on each compartment. The buoyancy chambers shall be so arranged that, in the event of any one of the compartments being damaged or failing to inflate, the intact compartments shall be able to support, with positive freeboard over the liferaft's entire periphery, the number of persons which the liferaft is permitted to accommodate, each having a mass of 75 kg and seated in their normal positions.

4.2.2.2 The floor of the liferaft shall be waterproof and shall be capable of being sufficiently insulated against cold either:

- **.1** by means of one or more compartments that the occupants can inflate, or which inflate automatically and can be deflated and re-inflated by the occupants; or
- **.2** by other equally efficient means not dependent on inflation.

4.2.2.3 The liferaft shall be capable of being inflated by one person. The liferaft shall be inflated with a non-toxic gas. The inflation system, including any relief valves installed in compliance with paragraph 4.2.2.4, shall comply with the requirements of an international standard acceptable to the Organization.* Inflation shall be completed within a period of 1 min at an ambient temperature of between 18°C and 20°C and within a period of 3 min at an ambient temperature of –30°C. After inflation, the liferaft shall maintain its form when loaded with its full complement of persons and equipment.

4.2.2.4 Each inflatable compartment shall be capable of withstanding a pressure equal to at least three times the working pressure and shall be prevented from reaching a pressure exceeding twice the working pressure either by means of relief valves or by a limited gas supply. Means shall be provided for fitting the topping-up pump or bellows required by paragraph 4.2.9.1.2 so that the working pressure can be maintained.

4.2.3 *Carrying capacity of inflatable liferafts*

The number of persons which a liferaft shall be permitted to accommodate shall be equal to the lesser of:

- **.1** the greatest whole number obtained by dividing by 0.096 the volume, measured in cubic metres, of the main buoyancy tubes (which for this purpose shall include neither the arches nor the thwarts, if fitted) when inflated; or

---

* Refer to the recommendations of the International Organization for Standardization, in particular publication ISO 15738: 2002 *Ships and marine technology – Gas inflation systems for inflatable life-saving appliances.*

.2 the greatest whole number obtained by dividing by 0.372 the inner horizontal cross-sectional area of the liferaft measured in square metres (which for this purpose may include the thwart or thwarts, if fitted) measured to the innermost edge of the buoyancy tubes; or

.3 the number of persons having an average mass of 75 kg, all wearing either immersion suits and lifejackets or, in the case of davit-launched liferafts, lifejackets, that can be seated with sufficient comfort and headroom without interfering with the operation of any of the liferaft's equipment.

4.2.4 *Access into inflatable liferafts*

4.2.4.1 At least one entrance shall be fitted with a boarding ramp, capable of supporting a person weighing 100 kg sitting or kneeling and not holding onto any other part of the liferaft, to enable persons to board the liferaft from the sea. The boarding ramp shall be so arranged as to prevent significant deflation of the liferaft if the ramp is damaged. In the case of a davit-launched liferaft having more than one entrance, the boarding ramp shall be fitted at the entrance opposite the bowsing lines and embarkation facilities.

4.2.4.2 Entrances not provided with a boarding ramp shall have a boarding ladder, the lowest step of which shall be situated not less than 0.4 m below the liferaft's light waterline.

4.2.4.3 There shall be means inside the liferaft to assist persons to pull themselves into the liferaft from the ladder.

4.2.5 *Stability of inflatable liferafts*

4.2.5.1 Every inflatable liferaft shall be so constructed that, when fully inflated and floating with the canopy uppermost, it is stable in a seaway.

4.2.5.2 The stability of the liferaft when in the inverted position shall be such that it can be righted in a seaway and in calm water by one person.

4.2.5.3 The stability of the liferaft when loaded with its full complement of persons and equipment shall be such that it can be towed at speeds of up to 3 knots in calm water.

4.2.5.4 The liferaft shall be fitted with water pockets complying with the following requirements:

.1 the water pockets shall be of a highly visible colour;

.2 the design shall be such that the pockets fill to at least 60% of their capacity within 25 s of deployment;

.3 the pockets shall have an aggregate capacity of at least 220 ℓ for liferafts up to 10 persons;

.4 the pockets for liferafts certified to carry more than 10 persons shall have an aggregate capacity of not less than 20*N* ℓ, where *N* = number of persons carried; and

.5 the pockets shall be positioned symmetrically round the circumference of the liferaft. Means shall be provided to enable air to readily escape from underneath the liferaft.

4.2.6 *Containers for inflatable liferafts*

4.2.6.1 The liferaft shall be packed in a container that is:

.1 so constructed as to withstand hard wear under conditions encountered at sea;

.2 of sufficient inherent buoyancy, when packed with the liferaft and its equipment, to pull the painter from within and to operate the inflation mechanism should the ship sink; and

.3 as far as practicable watertight, except for drain holes in the container bottom.

4.2.6.2 The liferaft shall be packed in its container in such a way as to ensure, as far as possible, that the waterborne liferaft inflates in an upright position on breaking free from its container.

4.2.6.3 The container shall be marked with:

.1 maker's name or trademark;

.2 serial number;

.3 name of approving authority and the number of persons it is permitted to carry;

.4 SOLAS;

.5 type of emergency pack enclosed;

.6 date when last serviced;

.7 length of painter;

.8 mass of the packed liferaft, if greater than 185 kg;

.9 maximum permitted height of stowage above waterline (depending on drop-test height and length of painter); and

.10 launching instructions.

4.2.7 *Markings on inflatable liferafts*

4.2.7.1 The liferaft shall be marked with:

.1 maker's name or trademark;

.2 serial number;

.3 date of manufacture (month and year);

.4 name of approving authority;

.5 name and place of servicing station where it was last serviced; and

.6 number of persons it is permitted to accommodate over each entrance in characters not less than 100 mm in height of a colour contrasting with that of the liferaft.

4.2.7.2 Provision shall be made for marking each liferaft with the name and port of registry of the ship to which it is to be fitted, in such a form that the ship identification can be changed at any time without opening the container.

4.2.8 *Davit-launched inflatable liferafts*

4.2.8.1 In addition to complying with the above requirements, a liferaft for use with an approved launching appliance shall, when suspended from its lifting hook or bridle, withstand a load of:

.1 four times the mass of its full complement of persons and equipment, at an ambient temperature and a stabilized liferaft temperature of 20 ± 3°C with all relief valves inoperative; and

.2 1.1 times the mass of its full complement of persons and equipment at an ambient temperature and a stabilized liferaft temperature of –30°C with all relief valves operative.

4.2.8.2 Rigid containers for liferafts to be launched by a launching appliance shall be so secured that the container or parts of it are prevented from falling into the sea during and after inflation and launching of the contained liferaft.

4.2.9 *Additional equipment for inflatable liferafts*

4.2.9.1 In addition to the equipment required by paragraph 4.1.5, every inflatable liferaft shall be provided with:

- **.1** one repair outfit for repairing punctures in buoyancy compartments; and
- **.2** one topping-up pump or bellows.

4.2.9.2 The knives required by paragraph 4.1.5.1.2 shall be safety knives, and the tin-openers and scissors required by paragraph 4.1.5.1.7 shall be of the safety type.

4.3 Rigid liferafts

4.3.1 Rigid liferafts shall comply with the requirements of section 4.1 and, in addition, shall comply with the requirements of this section.

4.3.2 *Construction of rigid liferafts*

4.3.2.1 The buoyancy of the liferaft shall be provided by approved inherently buoyant material placed as near as possible to the periphery of the liferaft. The buoyant material shall be fire-retardant or be protected by a fire-retardant covering.

4.3.2.2 The floor of the liferaft shall prevent the ingress of water and shall effectively support the occupants out of the water and insulate them from cold.

4.3.3 *Carrying capacity of rigid liferafts*

The number of persons which a liferaft shall be permitted to accommodate shall be equal to the lesser of:

- **.1** the greatest whole number obtained by dividing by 0.096 the volume, measured in cubic metres, of the buoyancy material multiplied by a factor of 1 minus the specific gravity of that material; or
- **.2** the greatest whole number obtained by dividing by 0.372 the horizontal cross-sectional area of the floor of the liferaft measured in square metres; or
- **.3** the number of persons having an average mass of 75 kg, all wearing immersion suits and lifejackets, that can be seated with sufficient comfort and headroom without interfering with the operation of any of the liferaft's equipment.

4.3.4 *Access into rigid liferafts*

4.3.4.1 At least one entrance shall be fitted with a boarding ramp, capable of supporting a person weighing 100 kg sitting or kneeling and not holding onto any other part of the liferaft, to enable persons to board the liferaft from the sea. In the case of a davit-launched liferaft having more than one entrance, the boarding ramp shall be fitted at the entrance opposite to the bowsing and embarkation facilities.

4.3.4.2 Entrances not provided with a boarding ramp shall have a boarding ladder, the lowest step of which shall be situated not less than 0.4 m below the liferaft's light waterline.

4.3.4.3 There shall be means inside the liferaft to assist persons to pull themselves into the liferaft from the ladder.

4.3.5 *Stability of rigid liferafts*

4.3.5.1 Unless the liferaft is capable of operating safely whichever way up it is floating, its strength and stability shall be such that it is either self-righting or can be readily righted in a seaway and in calm water by one person.

4.3.5.2 The stability of a liferaft when loaded with its full complement of persons and equipment shall be such that it can be towed at speeds of up to 3 knots in calm water.

4.3.6 *Markings on rigid liferafts*

The liferaft shall be marked with:

- **.1** name and port of registry of the ship to which it belongs;
- **.2** maker's name or trademark;
- **.3** serial number;
- **.4** name of approving authority;
- **.5** number of persons it is permitted to accommodate over each entrance in characters not less than 100 mm in height of a colour contrasting with that of the liferaft;
- **.6** SOLAS;
- **.7** type of emergency pack enclosed;
- **.8** length of painter;
- **.9** maximum permitted height of stowage above waterline (drop-test height); and
- **.10** launching instructions.

4.3.7 *Davit-launched rigid liferafts*

In addition to the above requirements, a rigid liferaft for use with an approved launching appliance shall, when suspended from its lifting hook or bridle, withstand a load of four times the mass of its full complement of persons and equipment.

4.4 General requirements for lifeboats

4.4.1 *Construction of lifeboats*

4.4.1.1 All lifeboats shall be properly constructed and shall be of such form and proportions that they have ample stability in a seaway and sufficient freeboard when loaded with their full complement of persons and equipment, and are capable of being safely launched under all conditions of trim of up to 10° and list of up to 20° either way. All lifeboats shall have rigid hulls and shall be capable of maintaining positive stability when in an upright position in calm water and loaded with their full complement of persons and equipment and holed in any one location below the waterline, assuming no loss of buoyancy material and no other damage.

4.4.1.2 Each lifeboat shall be fitted with a permanently affixed approval plate, endorsed by the Administration or its representative, containing at least the following items:

- **.1** manufacturer's name and address;
- **.2** lifeboat model and serial number;
- **.3** month and year of manufacture;
- **.4** number of persons the lifeboat is approved to carry; and
- **.5** the approval information required under paragraph 1.2.2.9.

Each production lifeboat shall be provided with a certificate or declaration of conformity which, in addition to the above items, specifies:

- **.6** number of the certificate of approval;
- **.7** material of hull construction, in such detail as to ensure that compatibility problems in repair should not occur;
- **.8** total mass fully equipped and fully manned;
- **.9** the measured towing force of the lifeboat; and
- **.10** statement of approval as to sections 4.5, 4.6, 4.7, 4.8 or 4.9.

4.4.1.3 All lifeboats shall be of sufficient strength to:

- **.1** enable them to be safely launched into the water when loaded with their full complement of persons and equipment; and

.2 be capable of being launched and towed when the ship is making headway at a speed of 5 knots in calm water.

4.4.1.4 Hulls and rigid covers shall be fire-retardant or non-combustible.

4.4.1.5 Seating shall be provided on thwarts, benches or fixed chairs which are constructed so as to be capable of supporting:

.1 a static load equivalent to the number of persons, each weighing 100 kg, for which spaces are provided in compliance with the requirements of paragraph 4.4.2.2.2;

.2 a load of 100 kg in any single seat location when a lifeboat to be launched by falls is dropped into the water from a height of at least 3 m; and

.3 a load of 100 kg in any single seat location when a free-fall lifeboat is launched from a height of at least 1.3 times its free-fall certification height.

4.4.1.6 Except for free-fall lifeboats, each lifeboat to be launched by falls shall be of sufficient strength to withstand a load, without residual deflection on removal of that load:

.1 in the case of boats with metal hulls, 1.25 times the total mass of the lifeboat when loaded with its full complement of persons and equipment; or

.2 in the case of other boats, twice the total mass of the lifeboat when loaded with its full complement of persons and equipment.

4.4.1.7 Except for free-fall lifeboats, each lifeboat to be launched by falls shall be of sufficient strength to withstand, when loaded with its full complement of persons and equipment and with, where applicable, skates or fenders in position, a lateral impact against the ship's side at an impact velocity of at least 3.5 m/s and also a drop into the water from a height of at least 3 m.

4.4.1.8 The vertical distance between the floor surface and the interior of the enclosure or canopy over 50% of the floor area shall be:

.1 not less than 1.3 m for a lifeboat permitted to accommodate 9 persons or less;

.2 not less than 1.7 m for a lifeboat permitted to accommodate 24 persons or more; and

.3 not less than the distance as determined by linear interpolation between 1.3 m and 1.7 m for a lifeboat permitted to accommodate between 9 and 24 persons.

4.4.2 *Carrying capacity of lifeboats*

4.4.2.1 No lifeboat shall be approved to accommodate more than 150 persons.

4.4.2.2 The number of persons which a lifeboat to be launched by falls shall be permitted to accommodate shall be equal to the lesser of:

- **.1** the number of persons having an average mass of 75 kg (for a lifeboat intended for a passenger ship) or 82.5 kg (for a lifeboat intended for a cargo ship), all wearing lifejackets, that can be seated in a normal position without interfering with the means of propulsion or the operation of any of the lifeboat's equipment; or
- **.2** the number of spaces that can be provided on the seating arrangements in accordance with figure 1. The shapes may be overlapped as shown, provided footrests are fitted and there is sufficient room for legs and the vertical separation between the upper and lower seat is not less than 350 mm.

4.4.2.3 Each seating position shall be clearly indicated in the lifeboat.

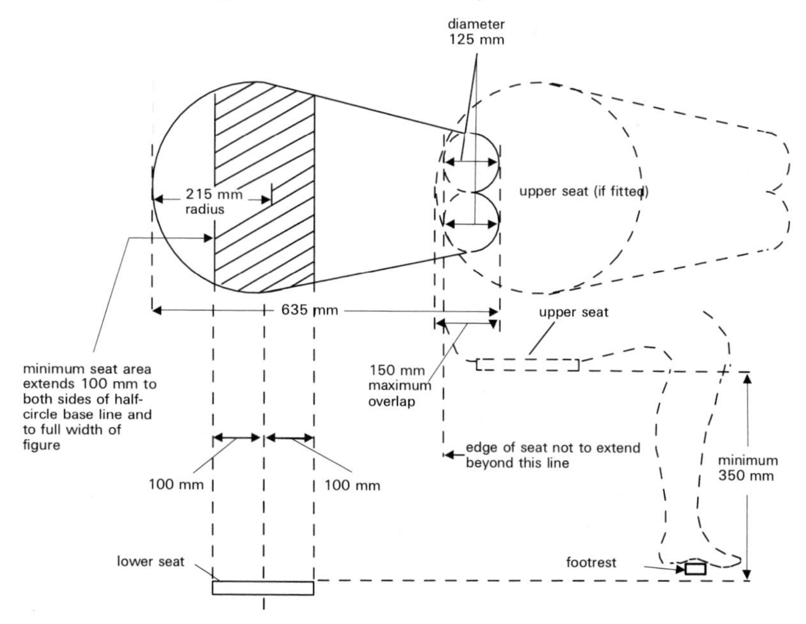

Figure 1

4.4.3 *Access into lifeboats*

4.4.3.1 Every passenger ship lifeboat shall be so arranged that it can be boarded by its full complement of persons in not more than 10 min from the time the instruction to board is given. Rapid disembarkation shall also be possible.

4.4.3.2 Every cargo ship lifeboat shall be so arranged that it can be boarded by its full complement of persons in not more than 3 min from the time the instruction to board is given. Rapid disembarkation shall also be possible.

4.4.3.3 Lifeboats shall have a boarding ladder that can be used at any boarding entrance of the lifeboat to enable persons in the water to board the lifeboat. The lowest step of the ladder shall be not less than 0.4 m below the lifeboat's light waterline.

4.4.3.4 The lifeboat shall be so arranged that helpless people can be brought on board either from the sea or on stretchers.

4.4.3.5 All surfaces on which persons might walk shall have a non-skid finish.

4.4.4 *Lifeboat buoyancy*

All lifeboats shall have inherent buoyancy or shall be fitted with inherently buoyant material which shall not be adversely affected by seawater, oil or oil products, sufficient to float the lifeboat with all its equipment on board when flooded and open to the sea. Additional inherently buoyant material, equal to 280 N of buoyant force per person, shall be provided for the number of persons the lifeboat is permitted to accommodate. Buoyant material, unless in addition to that required above, shall not be installed external to the hull of the lifeboat.

4.4.5 *Lifeboat freeboard and stability*

4.4.5.1 All lifeboats shall be stable and have a positive GM value when loaded with 50% of the number of persons the lifeboat is permitted to accommodate in their normal positions to one side of the centreline.

4.4.5.2 Under the condition of loading in paragraph 4.4.5.1:

- **.1** each lifeboat with side openings near the gunwale shall have a freeboard, measured from the waterline to the lowest opening through which the lifeboat may become flooded, of at least 1.5% of the lifeboat's length or 100 mm, whichever is the greater; and

.2 each lifeboat without side openings near the gunwale shall not exceed an angle of heel of 20° and shall have a freeboard, measured from the waterline to the lowest opening through which the lifeboat may become flooded, of at least 1.5% of the lifeboat's length or 100 mm, whichever is the greater.

4.4.6 *Lifeboat propulsion*

4.4.6.1 Every lifeboat shall be powered by a compression-ignition engine. No engine shall be used for any lifeboat if its fuel has a flashpoint of 43°C or less (closed-cup test).

4.4.6.2 The engine shall be provided with either a manual starting system, or a power starting system with two independent rechargeable energy sources. Any necessary starting aids shall also be provided. The engine starting systems and starting aids shall start the engine at an ambient temperature of −15°C within 2 min of commencing the start procedure unless, in the opinion of the Administration having regard to the particular voyages in which the ship carrying the lifeboat is constantly engaged, a different temperature is appropriate. The starting systems shall not be impeded by the engine casing, seating or other obstructions.

4.4.6.3 The engine shall be capable of operating for not less than 5 min after starting from cold with the lifeboat out of the water.

4.4.6.4 The engine shall be capable of operating when the lifeboat is flooded up to the centreline of the crankshaft.

4.4.6.5 The propeller shafting shall be so arranged that the propeller can be disengaged from the engine. Provision shall be made for ahead and astern propulsion of the lifeboat.

4.4.6.6 The exhaust pipe shall be so arranged as to prevent water from entering the engine in normal operation.

4.4.6.7 All lifeboats shall be designed with due regard to the safety of persons in the water and to the possibility of damage to the propulsion system by floating debris.

4.4.6.8 The speed of a lifeboat when proceeding ahead in calm water, when loaded with its full complement of persons and equipment and with all engine-powered auxiliary equipment in operation, shall be at least 6 knots and at least 2 knots when towing the largest liferaft carried on the ship, loaded with its full complement of persons and equipment or its equivalent. Sufficient fuel, suitable for use throughout the temperature range expected in the area in which the ship operates, shall be provided to run the fully loaded lifeboat at 6 knots for a period of not less than 24 h.

4.4.6.9 The lifeboat engine, transmission and engine accessories shall be enclosed in a fire-retardant casing or other suitable arrangements providing similar protection. Such arrangements shall also protect persons from coming into accidental contact with hot or moving parts and protect the engine from exposure to weather and sea. Adequate means shall be provided to reduce the engine noise so that a shouted order can be heard. Starter batteries shall be provided with casings which form a watertight enclosure around the bottom and sides of the batteries. The battery casings shall have a tightly fitting top which provides for necessary gas venting.

4.4.6.10 The lifeboat engine and accessories shall be designed to limit electromagnetic emissions so that engine operation does not interfere with the operation of radio life-saving appliances used in the lifeboat.

4.4.6.11 Means shall be provided for recharging all engine starting, radio and searchlight batteries. Radio batteries shall not be used to provide power for engine starting. Means shall be provided for recharging lifeboat batteries from the ship's power supply at a supply voltage not exceeding 50 V[*] which can be disconnected at the lifeboat embarkation station, or by means of a solar battery charger.

4.4.6.12 Water-resistant instructions for starting and operating the engine shall be provided and mounted in a conspicuous place near the engine starting controls.

4.4.7 *Lifeboat fittings*

4.4.7.1 All lifeboats except free-fall lifeboats shall be provided with at least one drain valve fitted near the lowest point in the hull, which shall automatically open to drain water from the hull when the lifeboat is not waterborne and shall automatically close to prevent entry of water when the lifeboat is waterborne. Each drain valve shall be provided with a cap or plug to close the valve, which shall be attached to the lifeboat by a lanyard, a chain, or other suitable means. Drain valves shall be readily accessible from inside the lifeboat and their position shall be clearly indicated.

4.4.7.2 All lifeboats shall be provided with a rudder and tiller. When a wheel or other remote steering mechanism is also provided the tiller shall be capable of controlling the rudder in case of failure of the steering mechanism.

[*] Refer to IEC 60921.

The rudder shall be permanently attached to the lifeboat. The tiller shall be permanently installed on, or linked to, the rudder stock; however, if the lifeboat has a remote steering mechanism, the tiller may be removable and securely stowed near the rudder stock. The rudder and tiller shall be so arranged as not to be damaged by operation of the release mechanism or the propeller.

4.4.7.3 Except in the vicinity of the rudder and propeller, suitable handholds shall be provided or a buoyant lifeline shall be becketed around the outside of the lifeboat above the waterline and within reach of a person in the water.

4.4.7.4 Lifeboats which are not self-righting when capsized shall have suitable handholds on the underside of the hull to enable persons to cling to the lifeboat. The handholds shall be fastened to the lifeboat in such a way that, when subjected to an impact sufficient to cause them to break away from the lifeboat, they break away without damaging the lifeboat.

4.4.7.5 All lifeboats shall be fitted with sufficient watertight lockers or compartments to provide for the storage of the small items of equipment, water and provisions required by paragraph 4.4.8. The lifeboat shall be equipped with a means for collecting rainwater, and in addition if required by the Administration a means for producing drinking water from seawater with a manually powered desalinator. The desalinator must not be dependent upon solar heat, nor on chemicals other than seawater. Means shall be provided for the storage of collected water.

4.4.7.6 Every lifeboat to be launched by a fall or falls, except a free-fall lifeboat, shall be fitted with a release mechanism complying with the following requirements subject to subparagraph .9 below:

.1 the mechanism shall be so arranged that all hooks are released simultaneously;

.2 the mechanism shall have two release capabilities: normal (off-load) release capability and on-load release capability:

.2.1 normal (off-load) release capability shall release the lifeboat when it is waterborne or when there is no load on the hooks, and not require manual separation of the lifting ring or shackle from the jaw of the hook; and

.2.2 on-load release capability shall release the lifeboat with a load on the hooks. This release shall be so arranged as to release the lifeboat under any conditions of loading from no load with the lifeboat waterborne to a load of 1.1 times the total mass of the lifeboat when loaded with its full complement of persons and equipment. This release capability shall be adequately protected against accidental or premature use. Adequate protection shall include special mechanical protection not normally required for off-load release, in addition to a danger sign. To prevent a premature on-load release, on-load operation of the release mechanism should require a deliberate and sustained action by the operator;

.3 to prevent an accidental release during recovery of the boat, unless the hook is completely reset, either the hook shall not be able to support any load, or the handle or safety pins shall not be able to be returned to the reset (closed) position without excessive force. Additional danger signs shall be posted at each hook station to alert crew members to the proper method of resetting;

.4 the release mechanism shall be so designed and installed that crew members from inside the lifeboat can clearly determine when the system is ready for lifting by:

.4.1 directly observing that the movable hook portion or the hook portion that locks the movable hook portion in place is properly and completely reset at each hook; or

.4.2 observing a non-adjustable indicator that confirms that the mechanism that locks the movable hook portion in place is properly and completely reset at each hook; or

.4.3 easily operating a mechanical indicator that confirms that the mechanism that locks the movable hook in place is properly and completely reset at each hook;

.5 clear operating instructions shall be provided with a suitably worded warning notice using colour coding, pictograms, and/or symbols as necessary for clarity. If colour coding is used, green shall indicate a properly reset hook and red shall indicate danger of improper or incorrect setting;

.6 the release control shall be clearly marked in a colour that contrasts with its surroundings;

.7 means shall be provided for hanging-off the lifeboat to free the release mechanism for maintenance;

.8 the fixed structural connections of the release mechanism in the lifeboat shall be designed with a calculated factor of safety of 6 based on the ultimate strength of the materials used, and the mass of the lifeboat when loaded with its full complement of persons, fuel and equipment, assuming the mass of the lifeboat is equally distributed between the falls, except that the factor of safety for the hanging-off arrangement may be based upon the mass of the lifeboat when loaded with its full complement of fuel and equipment plus 1,000 kg; and

.9 where a single fall and hook system is used for launching a lifeboat or rescue boat in combination with a suitable painter, the requirements of paragraphs 4.4.7.6.2.2 and 4.4.7.6.3 need not be applicable; in such an arrangement a single capability to release the lifeboat or rescue boat, only when it is fully waterborne, will be adequate.

4.4.7.7 Every lifeboat shall be fitted with a device to secure a painter near its bow. The device shall be such that the lifeboat does not exhibit unsafe or unstable characteristics when being towed by the ship making headway at speeds up to 5 knots in calm water. Except for free-fall lifeboats, the painter securing device shall include a release device to enable the painter to be released from inside the lifeboat, with the ship making headway at speeds up to 5 knots in calm water.

4.4.7.8 Every lifeboat which is fitted with a fixed two-way VHF radio-telephone apparatus with an antenna which is separately mounted shall be provided with arrangements for siting and securing the antenna effectively in its operating position.

4.4.7.9 Lifeboats intended for launching down the side of a ship shall have skates and fenders as necessary to facilitate launching and prevent damage to the lifeboat.

4.4.7.10 A manually controlled exterior light shall be fitted. The light shall be white and be capable of operating continuously for at least 12 h with a luminous intensity of not less than 4.3 cd in all directions of the upper hemisphere. However, if the light is a flashing light it shall flash at a rate of not less than 50 flashes and not more than 70 flashes per minute for the 12 h operating period with an equivalent effective luminous intensity.

4.4.7.11 A manually controlled exterior light or source of light shall be fitted inside the lifeboat to provide illumination for not less than 12 h to permit reading of survival and equipment instructions; however, oil lamps shall not be permitted for this purpose.

4.4.7.12 A manually controlled interior light shall be fitted inside the lifeboat capable of continuous operation for a period of at least 12 h. It shall produce an arithmetic mean luminous intensity of not less than 0.5 cd when measured over the entire upper hemisphere to permit reading of survival and equipment instructions; however, oil lamps shall not be permitted for this purpose.

4.4.8 *Lifeboat equipment*

All items of lifeboat equipment, whether required by this paragraph or elsewhere in section 4.4, shall be secured within the lifeboat by lashings, storage in lockers or compartments, storage in brackets or similar mounting arrangements or other suitable means. However, in the case of a lifeboat to be launched by falls, the boat-hooks shall be kept free for fending-off purposes. The equipment shall be secured in such a manner as not to interfere with any abandonment procedures. All items of lifeboat equipment shall be as small and of as little mass as possible and shall be packed in a suitable and compact form. Except where otherwise stated, the normal equipment of every lifeboat shall consist of:

.1 except for free-fall lifeboats, sufficient buoyant oars to make headway in calm seas. Thole pins, crutches or equivalent arrangements shall be provided for each oar provided. Thole pins or crutches shall be attached to the boat by lanyards or chains;

.2 two boat-hooks;

.3 a buoyant bailer and two buckets;

.4 a survival manual;*

.5 an operational compass which is luminous or provided with suitable means of illumination. In a totally enclosed lifeboat, the compass shall be permanently fitted at the steering position; in any other lifeboat, it shall be provided with a binnacle, if necessary to protect it from the weather, and suitable mounting arrangements;

.6 a sea-anchor of adequate size fitted with a shock-resistant hawser which provides a firm hand grip when wet. The strength of the sea-anchor, hawser and tripping line, if fitted, shall be adequate for all sea conditions;

.7 two efficient painters of a length equal to not less than twice the distance from the stowage position of the lifeboat to the waterline in the lightest seagoing condition or 15 m, whichever is the greater. On lifeboats to be launched by free-fall launching, both painters shall be stowed near the bow ready for use. On other lifeboats, one painter attached to the release device required by paragraph 4.4.7.7 shall be placed at the

* Refer to Instructions for action in survival craft, adopted by the Organization by resolution A.657(16).

forward end of the lifeboat and the other shall be firmly secured at or near the bow of the lifeboat ready for use;

.8 two hatchets, one at each end of the lifeboat;

.9 watertight receptacles containing a total of 3 ℓ of fresh water as described in paragraph 4.1.5.1.19 for each person the lifeboat is permitted to accommodate, of which either 1 ℓ per person may be replaced by a desalting apparatus capable of producing an equal amount of fresh water in two days or 2 ℓ per person may be replaced by a manually powered reverse-osmosis desalinator, as described in paragraph 4.4.7.5, capable of producing an equal amount of fresh water in two days;

.10 a rustproof dipper with lanyard;

.11 a rustproof graduated drinking vessel;

.12 a food ration as described in paragraph 4.1.5.1.18 totalling not less than 10,000 kJ for each person the lifeboat is permitted to accommodate; these rations shall be kept in airtight packaging and be stowed in a watertight container;

.13 four rocket parachute flares complying with the requirements of section 3.1;

.14 six hand flares complying with the requirements of section 3.2;

.15 two buoyant smoke signals complying with the requirements of section 3.3;

.16 one waterproof electric torch suitable for Morse signalling together with one spare set of batteries and one spare bulb in a waterproof container;

.17 one daylight signalling mirror with instructions for its use for signalling to ships and aircraft;

.18 one copy of the life-saving signals prescribed by regulation V/29 on a waterproof card or in a waterproof container;

.19 one whistle or equivalent sound signal;

.20 a first-aid outfit in a waterproof case capable of being closed tightly after use;

.21 anti-seasickness medicine sufficient for at least 48 h and one seasickness bag for each person;

.22 a jack-knife, to be kept attached to the boat by a lanyard;

.23 three tin-openers;

.24 two buoyant rescue quoits, attached to not less than 30 m of buoyant line;

.25 if the lifeboat is not automatically self-bailing, a manual pump suitable for effective bailing;

.26 one set of fishing tackle;

.27 sufficient tools for minor adjustments to the engine and its accessories;

.28 portable fire-extinguishing equipment of an approved type suitable for extinguishing oil fires;*

.29 a searchlight with a horizontal and vertical sector of at least 6° and a measured luminous intensity of 2,500 cd which can work continuously for not less than 3 h;

.30 an efficient radar reflector, unless a survival craft radar transponder is stowed in the lifeboat;

.31 thermal protective aids complying with the requirements of section 2.5 sufficient for 10% of the number of persons the lifeboat is permitted to accommodate or two, whichever is the greater; and

.32 in the case of ships engaged on voyages of such a nature and duration that, in the opinion of the Administration, the items specified in paragraphs 4.4.8.12 and 4.4.8.26 are unnecessary, the Administration may allow these items to be dispensed with.

4.4.9 *Lifeboat markings*

4.4.9.1 The number(s) of persons for which the lifeboat is approved, for passenger ships and/or cargo ships, as applicable, shall be clearly marked on it in clear permanent characters.

4.4.9.2 The name and port of registry of the ship to which the lifeboat belongs shall be marked on each side of the lifeboat's bow in block capitals of the Roman alphabet.

4.4.9.3 Means of identifying the ship to which the lifeboat belongs and the number of the lifeboat shall be marked in such a way that they are visible from above.

* Refer to the Improved Guidelines for marine portable fire extinguishers, adopted by the Organization by resolution A.951(23).

4.5 Partially enclosed lifeboats

4.5.1 Partially enclosed lifeboats shall comply with the requirements of section 4.4 and in addition shall comply with the requirements of this section.

4.5.2 Partially enclosed lifeboats shall be provided with permanently attached rigid covers extending over not less than 20% of the length of the lifeboat from the stem and not less than 20% of the length of the lifeboat from the aftermost part of the lifeboat. The lifeboat shall be fitted with a permanently attached foldable canopy which together with the rigid covers completely encloses the occupants of the lifeboat in a weatherproof shelter and protects them from exposure. The lifeboat shall have entrances at both ends and on each side. Entrances in the rigid covers shall be weathertight when closed. The canopy shall be so arranged that:

- **.1** it is provided with adequate rigid sections or battens to permit erection of the canopy;
- **.2** it can be easily erected by not more than two persons;
- **.3** it is insulated to protect the occupants against heat and cold by means of not less than two layers of material separated by an air gap or other equally efficient means; means shall be provided to prevent accumulation of water in the air gap;
- **.4** its exterior is of a highly visible colour and its interior is of a colour which does not cause discomfort to the occupants;
- **.5** entrances in the canopy are provided with efficient adjustable closing arrangements which can be easily and quickly opened and closed from inside or outside so as to permit ventilation but exclude seawater, wind and cold; means shall be provided for holding the entrances securely in the open and closed position;
- **.6** with the entrances closed, it admits sufficient air for the occupants at all times;
- **.7** it has means for collecting rainwater; and
- **.8** the occupants can escape in the event of the lifeboat capsizing.

4.5.3 The interior of the lifeboat shall be of a light colour which does not cause discomfort to the occupants.

4.5.4 If a fixed two-way VHF radiotelephone apparatus is fitted in the lifeboat, it shall be installed in a cabin large enough to accommodate both the equipment and the person using it. No separate cabin is required if the construction of the lifeboat provides a sheltered space to the satisfaction of the Administration.

4.6 Totally enclosed lifeboats

4.6.1 Totally enclosed lifeboats shall comply with the requirements of section 4.4 and in addition shall comply with the requirements of this section.

4.6.2 *Enclosure*

Every totally enclosed lifeboat shall be provided with a rigid watertight enclosure which completely encloses the lifeboat. The enclosure shall be so arranged that:

.1 it provides shelter for the occupants;

.2 access to the lifeboat is provided by hatches which can be closed to make the lifeboat watertight;

.3 except for free-fall lifeboats, hatches are positioned so as to allow launching and recovery operations to be performed without any occupant having to leave the enclosure;

.4 access hatches are capable of being opened and closed from both inside and outside and are equipped with means to hold them securely in open positions;

.5 except for a free-fall lifeboat, it is possible to row the lifeboat;

.6 it is capable, when the lifeboat is in the capsized position with the hatches closed and without significant leakage, of supporting the entire mass of the lifeboat, including all equipment, machinery and its full complement of persons;

.7 it includes windows or translucent panels which admit sufficient daylight to the inside of the lifeboat with the hatches closed to make artificial light unnecessary;

.8 its exterior is of a highly visible colour and its interior of a light colour which does not cause discomfort to the occupants;

.9 handrails provide a secure handhold for persons moving about the exterior of the lifeboat, and aid embarkation and disembarkation;

.10 persons have access to their seats from an entrance without having to climb over thwarts or other obstructions; and

.11 during operation of the engine with the enclosure closed, the atmospheric pressure inside the lifeboat shall never be above or below the outside atmospheric pressure by more than 20 hPa.

4.6.3 *Capsizing and re-righting*

4.6.3.1 Except in free-fall lifeboats, a safety belt shall be fitted at each indicated seating position. The safety belt shall be designed to hold a person with a mass of 100 kg securely in place when the lifeboat is in a capsized position. Each set of safety belts for a seat shall be of a colour which contrasts with the belts for seats immediately adjacent. Free-fall lifeboats shall be fitted with a safety harness at each seat in contrasting colour designed to hold a person with a mass of 100 kg securely in place during a free-fall launch as well as with the lifeboat in capsized position.

4.6.3.2 The stability of the lifeboat shall be such that it is inherently or automatically self-righting when loaded with its full or a partial complement of persons and equipment and all entrances and openings are closed watertight and the persons are secured with safety belts.

4.6.3.3 The lifeboat shall be capable of supporting its full complement of persons and equipment when the lifeboat is in the damaged condition prescribed in paragraph 4.4.1.1 and its stability shall be such that, in the event of capsizing, it will automatically attain a position that will provide an above-water escape for its occupants. When the lifeboat is in the stable flooded condition, the water level inside the lifeboat, measured along the seat back, shall not be more than 500 mm above the seat pan at any occupant seating position.

4.6.3.4 The design of all engine exhaust pipes, air ducts and other openings shall be such that water is excluded from the engine when the lifeboat capsizes and re-rights.

4.6.4 *Propulsion*

4.6.4.1 The engine and transmission shall be controlled from the helmsman's position.

4.6.4.2 The engine and engine installation shall be capable of running in any position during capsize and continue to run after the lifeboat returns to the upright or shall automatically stop on capsizing and be easily restarted after the lifeboat returns to the upright. The design of the fuel and lubricating systems shall prevent the loss of fuel and the loss of more than 250 mℓ of lubricating oil from the engine during capsize.

4.6.4.3 Air-cooled engines shall have a duct system to take in cooling air from, and exhaust it to, the outside of the lifeboat. Manually operated dampers shall be provided to enable cooling air to be taken in from, and exhausted to, the interior of the lifeboat.

4.6.5 *Protection against acceleration*

Notwithstanding paragraph 4.4.1.7, a totally enclosed lifeboat, except a free-fall lifeboat, shall be so constructed and fendered that the lifeboat renders protection against harmful accelerations resulting from an impact of the lifeboat, when loaded with its full complement of persons and equipment, against the ship's side at an impact velocity of not less than 3.5 m/s.

4.7 Free-fall lifeboats

4.7.1 *General requirements*

Free-fall lifeboats shall comply with the requirements of section 4.6 and in addition shall comply with the requirements of this section.

4.7.2 *Carrying capacity of a free-fall lifeboat*

4.7.2.1 The carrying capacity of a free fall lifeboat is the number of persons having an average mass of 82.5 kg that can be provided with a seat without interfering with the means of propulsion or the operation of any of the lifeboat's equipment. The seating surface shall be smooth and shaped and provided with cushioning of at least 10 mm over all contact areas to provide support for the back and pelvis and flexible lateral side support for the head. The seats shall be of the non-folding type, permanently secured to the lifeboat and arranged so that any deflection of the hull or canopy during launching will not cause injury to the occupants. The location and structure of the seat shall be arranged to preclude the potential for injury during launch if the seat is narrower than the occupant's shoulders. The passage between the seats shall have a clear width of at least 480 mm from the deck to the top of the seats, be free of any obstruction and provided with an antislip surface with suitable footholds to allow safe embarkation in the ready-to-launch position. Each seat shall be provided with a suitable locking harness capable of quick release under tension to restrain the body of the occupant during launching.

4.7.2.2 The angle between the seat pan and the seat back shall be at least 90°. The width of the seat pan shall be at least 480 mm. Free clearance in front of the backrest (buttock to knee length) shall be at least 650 mm measured at an angle of 90° to the backrest.

The backrest shall extend at least 1,075 mm above the seat pan. The seat shall provide for shoulder height, measured along the seat back, of at least 760 mm. The footrest shall be oriented at not less than half of the angle of the seat pan and shall have a foot length of at least 330 mm (see figure 2).

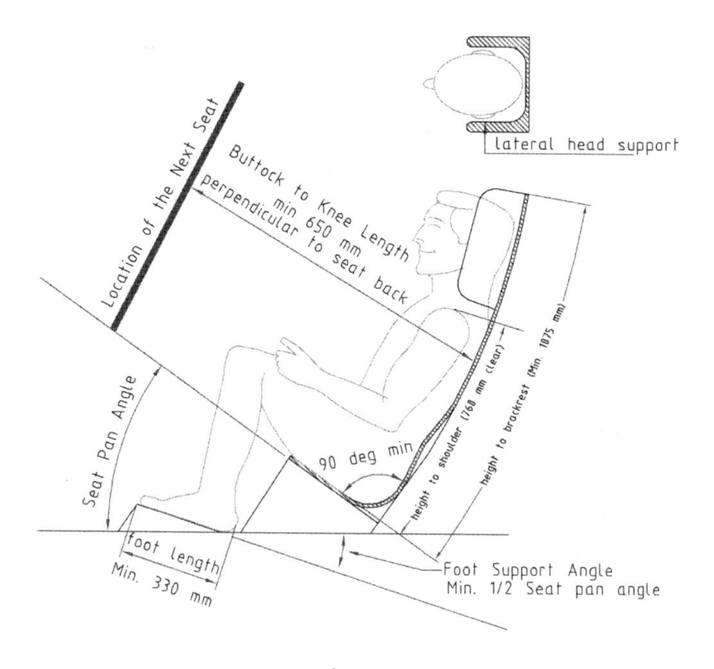

Figure 2

4.7.3 *Performance requirements*

4.7.3.1 Each free-fall lifeboat shall make positive headway immediately after water entry and shall not come into contact with the ship after a free-fall launching against a trim of up to 10° and a list of up to 20° either way from the certification height when fully equipped and loaded with:

.1 its full complement of persons;

.2 occupants so as to cause the centre of gravity to be in the most forward position;

.3 occupants so as to cause the centre of gravity to be in the most aft position; and

.4 its operating crew only.

4.7.3.2 For oil tankers, chemical tankers and gas carriers with a final angle of heel greater than 20° calculated in accordance with the International Convention for the Prevention of Pollution from Ships, 1973, as modified by the Protocol of 1978 relating thereto, and the recommendations of the Organization,* as applicable, a lifeboat shall be capable of being free-fall launched at the final angle of heel and on the base of the final waterline of that calculation.

4.7.4 *Construction*

Each free-fall lifeboat shall be of sufficient strength to withstand, when loaded with its full complement of persons and equipment, a free-fall launch from a height of at least 1.3 times the free-fall certification height.

4.7.5 *Protection against harmful acceleration*

Each free-fall lifeboat shall be so constructed as to ensure that the lifeboat is capable of rendering protection against harmful accelerations resulting from being launched from the height for which it is to be certified in calm water under unfavourable conditions of trim of up to 10° and list of up to 20° either way when it is fully equipped and loaded with:

.1 its full complement of persons;

.2 occupants so as to cause the centre of gravity to be in the most forward position;

.3 occupants so as to cause the centre of gravity to be in the most aft position; and

.4 the operating crew only.

4.7.6 *Lifeboat fittings*

Each free-fall lifeboat shall be fitted with a release system which shall:

.1 have two independent activation systems for the release mechanisms which may only be operated from inside the lifeboat and be marked in a colour that contrasts with its surroundings;

---

* Refer to the damage stability requirements of the *International Code for the Construction and Equipment of Ships Carrying Dangerous Chemicals in Bulk (IBC Code)*, adopted by the Maritime Safety Committee by resolution MSC.4(48) and the *International Code for the Construction and Equipment of Ships Carrying Liquefied Gases in Bulk (IGC Code)*, adopted by the Maritime Safety Committee by resolution MSC.5(48).

.2 be so arranged as to release the boat under any condition of loading from no load up to at least 200% of the normal load caused by the fully equipped lifeboat when loaded with the number of persons for which it is to be approved;

.3 be adequately protected against accidental or premature use;

.4 be designed to test the release system without launching the lifeboat; and

.5 be designed with a factor of safety of 6 based on the ultimate strength of the materials used.

4.7.7 *Certificate of approval*

In addition to the requirements of paragraph 4.4.1.2, the certificate of approval for a free-fall lifeboat shall also state:

.1 free-fall certification height;

.2 required launching ramp length; and

.3 launching ramp angle for the free-fall certification height.

4.8 Lifeboats with a self-contained air support system

In addition to complying with the requirements of section 4.6 or 4.7, as applicable, a lifeboat with a self-contained air support system shall be so arranged that, when proceeding with all entrances and openings closed, the air in the lifeboat remains safe and breathable and the engine runs normally for a period of not less than 10 min. During this period the atmospheric pressure inside the lifeboat shall never fall below the outside atmospheric pressure nor shall it exceed it by more than 20 hPa. The system shall have visual indicators to indicate the pressure of the air supply at all times.

4.9 Fire-protected lifeboats

4.9.1 In addition to complying with the requirements of section 4.8, a fire-protected lifeboat, when waterborne, shall be capable of protecting the number of persons it is permitted to accommodate when subjected to a continuous oil fire that envelops the lifeboat for a period of not less than 8 min.

4.9.2 *Water spray system*

A lifeboat which has a water spray fire-protection system shall comply with the following:

.1 water for the system shall be drawn from the sea by a self-priming motor pump. It shall be possible to turn "on" and turn "off" the flow of water over the exterior of the lifeboat;

.2 the seawater intake shall be so arranged as to prevent the intake of flammable liquids from the sea surface; and

.3 the system shall be arranged for flushing with fresh water and allowing complete drainage.

Chapter V
Rescue boats

5.1 Rescue boats

5.1.1 *General requirements*

5.1.1.1 Except as provided by this section, all rescue boats shall comply with the requirements of paragraphs 4.4.1 to 4.4.7.4 inclusive, excluding paragraphs 4.4.6.8, 4.4.7.6, 4.4.7.8, 4.4.7.10, 4.4.7.11 and 4.4.9, except that, for all rescue boats, an average mass of 82.5 kg shall apply to paragraph 4.4.2.2.1. A lifeboat may be approved and used as a rescue boat if it meets all of the requirements of this section, if it successfully completes the testing for a rescue boat required in regulation III/4.2, and if its stowage, launching and recovery arrangements on the ship meet all of the requirements for a rescue boat.

5.1.1.2 Notwithstanding the requirements of paragraph 4.4.4, required buoyant material for rescue boats may be installed external to the hull, provided it is adequately protected against damage and is capable of withstanding exposure as specified in paragraph 5.1.3.3.

5.1.1.3 Rescue boats may be either of rigid or inflated construction or a combination of both and shall:

.1 be not less than 3.8 m and not more than 8.5 m in length; and

.2 be capable of carrying at least five seated persons and a person lying on a stretcher all wearing immersion suits and lifejackets if required. Notwithstanding paragraph 4.4.1.5, seating, except for the helmsman, may be provided on the floor, provided that the seating space analysis in accordance with paragraph 4.4.2.2.2 uses shapes similar to figure 1, but altered to an overall length of 1,190 mm to provide for extended legs. No part of a seating space shall be on the gunwale, transom, or on inflated buoyancy at the sides of the boat.

5.1.1.4 Rescue boats which are a combination of rigid and inflated construction shall comply with the appropriate requirements of this section to the satisfaction of the Administration.

5.1.1.5 Unless the rescue boat has adequate sheer, it shall be provided with a bow cover extending for not less than 15% of its length.

5.1.1.6 Every rescue boat shall be provided with sufficient fuel, suitable for use throughout the temperature range expected in the area in which the ship operates, and be capable of manoeuvring at a speed of at least 6 knots and maintaining that speed, for a period of at least 4 h, when loaded with its full complement of persons and equipment.

5.1.1.7 Rescue boats shall have sufficient mobility and manoeuvrability in a seaway to enable persons to be retrieved from the water, marshal liferafts and tow the largest liferaft carried on the ship when loaded with its full complement of persons and equipment or its equivalent at a speed of at least 2 knots.

5.1.1.8 A rescue boat shall be fitted with an inboard engine or outboard motor. If it is fitted with an outboard motor, the rudder and tiller may form part of the engine. Notwithstanding the requirements of paragraph 4.4.6.1, petrol-driven outboard engines with an approved fuel system may be fitted in rescue boats provided the fuel tanks are specially protected against fire and explosion.

5.1.1.9 Arrangements for towing shall be permanently fitted in rescue boats and shall be sufficiently strong to marshal or tow liferafts as required by paragraph 5.1.1.7.

5.1.1.10 Unless expressly provided otherwise, every rescue boat shall be provided with effective means of bailing or be automatically self-bailing.

5.1.1.11 Rescue boats shall be fitted with weathertight stowage for small items of equipment.

5.1.1.12 Every rescue boat shall be so arranged that an adequate view forward, aft and to both sides is provided from the control and steering position for safe launching and manoeuvring and, in particular, with regard to visibility of areas and crew members essential to man-overboard retrieval and marshalling of survival craft.

5.1.2 *Rescue boat equipment*

5.1.2.1 All items of rescue boat equipment, with the exception of boat-hooks which shall be kept free for fending-off purposes, shall be secured within the rescue boat by lashings, storage in lockers or compartments, storage in brackets or similar mounting arrangements, or other suitable means. The equipment shall be secured in such a manner as not to interfere with any launching or recovery procedures. All items of rescue boat equipment shall be as small and of as little mass as possible and shall be packed in suitable and compact form.

5.1.2.2 The normal equipment of every rescue boat shall consist of:

.1 sufficient buoyant oars or paddles to make headway in calm seas. Thole pins, crutches or equivalent arrangements shall be provided for each oar. Thole pins or crutches shall be attached to the boat by lanyards or chains;

.2 a buoyant bailer;

.3 a binnacle containing an efficient compass which is luminous or provided with suitable means of illumination;

.4 a sea-anchor and tripping line, if fitted, with a hawser of adequate strength not less than 10 m in length;

.5 a painter of sufficient length and strength, attached to the release device complying with the requirements of paragraph 4.4.7.7 and placed at the forward end of the rescue boat;

.6 one buoyant line, not less than 50 m in length, of sufficient strength to tow a liferaft as required by paragraph 5.1.1.7;

.7 one waterproof electric torch suitable for Morse signalling, together with one spare set of batteries and one spare bulb in a waterproof container;

.8 one whistle or equivalent sound signal;

.9 a first-aid outfit in a waterproof case capable of being closed tightly after use;

.10 two buoyant rescue quoits, attached to not less than 30 m of buoyant line;

.11 a searchlight with a horizontal and vertical sector of at least 6° and a measured luminous intensity of 2,500 cd which can work continuously for not less than 3 h;

.12 an efficient radar reflector;

.13 thermal protective aids complying with the requirements of section 2.5 sufficient for 10% of the number of persons the rescue boat is permitted to accommodate or two, whichever is the greater; and

.14 portable fire-extinguishing equipment of an approved type suitable for extinguishing oil fires.*

5.1.2.3 In addition to the equipment required by paragraph 5.1.2.2, the normal equipment of every rigid rescue boat shall include:

.1 a boat-hook;

.2 a bucket; and

.3 a knife or hatchet.

5.1.2.4 In addition to the equipment required by paragraph 5.1.2.2, the normal equipment of every inflated rescue boat shall consist of:

.1 a buoyant safety knife;

.2 two sponges;

.3 an efficient manually operated bellows or pump;

.4 a repair kit in a suitable container for repairing punctures; and

.5 a safety boat-hook.

5.1.3 *Additional requirements for inflated rescue boats*

5.1.3.1 The requirements of paragraphs 4.4.1.4 and 4.4.1.6 do not apply to inflated rescue boats.

5.1.3.2 An inflated rescue boat shall be constructed in such a way that, when suspended by its bridle or lifting hook:

.1 it is of sufficient strength and rigidity to enable it to be lowered and recovered with its full complement of persons and equipment;

.2 it is of sufficient strength to withstand a load of four times the mass of its full complement of persons and equipment at an ambient temperature of 20 ± 3°C, with all relief valves inoperative; and

.3 it is of sufficient strength to withstand a load of 1.1 times the mass of its full complement of persons and equipment at an ambient temperature of –30°C, with all relief valves operative.

---

* Refer to the Improved Guidelines for marine portable fire extinguishers, adopted by the Organization by resolution A.951(23).

5.1.3.3 Inflated rescue boats shall be so constructed as to be capable of withstanding exposure:

- **.1** when stowed on an open deck on a ship at sea;
- **.2** for 30 days afloat in all sea conditions.

5.1.3.4 In addition to complying with the requirements of paragraph 4.4.9, inflated rescue boats shall be marked with a serial number, the maker's name or trademark and the date of manufacture.

5.1.3.5 The buoyancy of an inflated rescue boat shall be provided by either a single tube subdivided into at least five separate compartments of approximately equal volume or two separate tubes, neither exceeding 60% of the total volume. The buoyancy tubes shall be so arranged that the intact compartments shall be able to support the number of persons which the rescue boat is permitted to accommodate, each having a mass of 82.5 kg, when seated in their normal positions with positive freeboard over the rescue boat's entire periphery under the following conditions:

- **.1** with the forward buoyancy compartment deflated;
- **.2** with the entire buoyancy on one side of the rescue boat deflated; and
- **.3** with the entire buoyancy on one side and the bow compartment deflated.

5.1.3.6 The buoyancy tubes forming the boundary of the inflated rescue boat shall, on inflation, provide a volume of not less than 0.17 m^3 for each person the rescue boat is permitted to accommodate.

5.1.3.7 Each buoyancy compartment shall be fitted with a nonreturn valve for manual inflation and means for deflation. A safety relief valve shall also be fitted unless the Administration is satisfied that such an appliance is unnecessary.

5.1.3.8 Underneath the bottom and on vulnerable places on the outside of the inflated rescue boat, rubbing strips shall be provided to the satisfaction of the Administration.

5.1.3.9 Where a transom is fitted it shall not be inset by more than 20% of the overall length of the rescue boat.

5.1.3.10 Suitable patches shall be provided for securing the painters fore and aft and the becketed lifelines inside and outside the boat.

5.1.4 *Additional requirements for fast rescue boats*

5.1.4.1 Fast rescue boats shall be so constructed to be capable of being safely launched and retrieved under adverse weather and sea conditions.

5.1.4.2 Except as provided by this section, all fast rescue boats shall comply with the requirements of section 5.1, except for paragraphs 4.4.1.5.3, 4.4.1.6, 4.4.7.2, 5.1.1.6 and 5.1.1.10.

5.1.4.3 Notwithstanding paragraph 5.1.1.3.1, fast rescue boats shall have a hull length of not less than 6 m and not more than 8.5 m, including inflated structures or fixed fenders.

5.1.4.4 Fast rescue boats shall be provided with sufficient fuel, suitable for use throughout the temperature range expected in the area in which the ship operates, and be capable of manoeuvring, for a period of at least 4 h, at a speed of at least 20 knots in calm water with a crew of 3 persons and at least 8 knots when loaded with its full complement of persons and equipment.

5.1.4.5 Fast rescue boats shall be self-righting or capable of being readily righted by not more than two of their crew.

5.1.4.6 Fast rescue boats shall be self-bailing or be capable of being rapidly cleared of water.

5.1.4.7 Fast rescue boats shall be steered by a wheel at the helmsman's position remote from the tiller. An emergency steering system providing direct control of the rudder, water jet, or outboard motor shall also be provided.

5.1.4.8 Engines in fast rescue boats shall stop automatically or be stopped by the helmsman's emergency release switch, should the rescue boat capsize. When the rescue boat has righted, each engine or motor shall be capable of being restarted provided that the helmsman's emergency release, if fitted, has been reset. The design of the fuel and lubricating systems shall prevent the loss of more than 250 ml of fuel or lubricating oil from the propulsion system, should the rescue boat capsize.

5.1.4.9 Fast rescue boats shall, if possible, be equipped with an easily and safely operated fixed single-point suspension arrangement or equivalent.

5.1.4.10 A rigid fast rescue boat shall be constructed in such a way that, when suspended by its lifting point, it is of sufficient strength to withstand a load of 4 times the mass of its full complement of persons and equipment without residual deflection upon removal of the load.

5.1.4.11 The normal equipment of a fast rescue boat shall include a VHF radiocommunication set which is hands-free and watertight.

Chapter VI

Launching and embarkation appliances

6.1 Launching and embarkation appliances

6.1.1 *General requirements*

6.1.1.1 With the exception of the secondary means of launching for free-fall lifeboats, each launching appliance shall be so arranged that the fully equipped survival craft or rescue boat it serves can be safely launched against unfavourable conditions of trim of up to 10° and a list of up to 20° either way:

.1 when boarded, as required by regulation III/23 or III/33, by its full complement of persons; and

.2 with not more than the required operating crew on board.

6.1.1.2 Notwithstanding the requirements of paragraph 6.1.1.1, lifeboat launching appliances for oil tankers, chemical tankers and gas carriers with a final angle of heel greater than 20° calculated in accordance with the International Convention for the Prevention of Pollution from Ships, 1973, as modified by the Protocol of 1978 relating thereto, and the recommendations of the Organization,* as applicable, shall be capable of operating at the final angle of heel on the lower side of the ship, taking into consideration the final damaged waterline of the ship.

6.1.1.3 A launching appliance shall not depend on any means other than gravity or stored mechanical power which is independent of the ship's power supplies to launch the survival craft or rescue boat it serves in the fully loaded and equipped condition and also in the light condition.

6.1.1.4 Each launching appliance shall be so constructed that only a minimum amount of routine maintenance is necessary. All parts requiring

* Refer to the damage stability requirements of the *International Code for the Construction and Equipment of Ships Carrying Dangerous Chemicals in Bulk (IBC Code)*, adopted by the Maritime Safety Committee by resolution MSC.4(48) and the *International Code for the Construction and Equipment of Ships Carrying Liquefied Gases in Bulk (IGC Code)*, adopted by the Maritime Safety Committee by resolution MSC.5(48).

regular maintenance by the ship's crew shall be readily accessible and easily maintained.

6.1.1.5 The launching appliance and its attachments other than winch brakes shall be of sufficient strength to withstand a factory static proof load test of not less than 2.2 times the maximum working load.

6.1.1.6 Structural members and all blocks, falls, padeyes, links, fastenings and all other fittings used in connection with launching equipment shall be designed with a factor of safety on the basis of the maximum working load assigned and the ultimate strengths of the materials used for construction. A minimum factor of safety of 4.5 shall be applied to all structural members, and a minimum factor of safety of 6 shall be applied to falls, suspension chains, links and blocks.

6.1.1.7 Each launching appliance shall, as far as practicable, remain effective under conditions of icing.

6.1.1.8 A lifeboat launching appliance shall be capable of recovering the lifeboat with its crew.

6.1.1.9 Each rescue boat launching appliance shall be fitted with a powered winch motor capable of raising the rescue boat from the water with its full rescue boat complement of persons and equipment at a rate of not less than 0.3 m/s.

6.1.1.10 The arrangements of the launching appliance shall be such as to enable safe boarding of the survival craft in accordance with the requirements of paragraphs 4.1.4.2, 4.1.4.3, 4.4.3.1 and 4.4.3.2.

6.1.1.11 Rescue boat launching appliances shall be provided with foul weather recovery strops for recovery where heavy fall blocks constitute a danger.

6.1.2 *Launching appliances using falls and a winch*

6.1.2.1 Every launching appliance using falls and a winch, except for secondary launching appliances for free-fall lifeboats, shall comply with the requirements of paragraph 6.1.1 and, in addition, shall comply with the requirements of this paragraph.

6.1.2.2 The launching mechanism shall be so arranged that it may be actuated by one person from a position on the ship's deck and, except for secondary launching appliances for free-fall lifeboats, from a position within the survival craft or rescue boat. When launched by a person on the deck, the survival craft or rescue boat shall be visible to that person.

6.1.2.3 Falls shall be of rotation-resistant and corrosion-resistant steel wire rope.

6.1.2.4 In the case of a multiple-drum winch, unless an efficient compensatory device is fitted, the falls shall be so arranged as to wind off the drums at the same rate when lowering, and to wind on to the drums evenly at the same rate when hoisting.

6.1.2.5 The winch brakes of a launching appliance shall be of sufficient strength to withstand:

.1 a static test with a proof load of not less than 1.5 times the maximum working load; and

.2 a dynamic test with a proof load of not less than 1.1 times the maximum working load at maximum lowering speed.

6.1.2.6 An efficient hand gear shall be provided for recovery of each survival craft and rescue boat. Hand-gear handles or wheels shall not be rotated by moving parts of the winch when the survival craft or rescue boat is being lowered or when it is being hoisted by power.

6.1.2.7 Where davit arms are recovered by power, safety devices shall be fitted which will automatically cut off the power before the davit arms reach the stops in order to prevent overstressing the falls or davits, unless the motor is designed to prevent such overstressing.

6.1.2.8 The speed at which the fully loaded survival craft or rescue boat is lowered to the water shall not be less than that obtained from the formula:

$$S = 0.4 + 0.02\,H$$

where:

S is the lowering speed in metres per second and

H is the height in metres from the davit head to the waterline with the ship in the lightest sea-going condition.

6.1.2.9 The lowering speed of a fully equipped liferaft without persons on board shall be to the satisfaction of the Administration. The lowering speed of other survival craft, fully equipped but without persons on board, shall be at least 70% of that required by paragraph 6.1.2.8.

6.1.2.10 The maximum lowering speed shall be established by the Administration having regard to the design of the survival craft or rescue boat, the protection of its occupants from excessive forces, and the strength of the launching arrangements taking into account inertia forces during an emergency stop. Means shall be incorporated into the appliance to ensure that this speed is not exceeded.

6.1.2.11 Every launching appliance shall be fitted with brakes capable of stopping the descent of the survival craft or rescue boat and holding it securely when loaded with its full complement of persons and equipment; brake pads shall, where necessary, be protected from water and oil.

6.1.2.12 Manual brakes shall be so arranged that the brake is always applied unless the operator, either on deck or in the survival craft or rescue boat, holds the brake control in the "off" position.

6.1.2.13 A lifeboat launching appliance shall be provided with means for hanging-off the lifeboat to free the on-load release mechanism for maintenance.

6.1.3 *Float-free launching*

Where a survival craft requires a launching appliance and is also designed to float free, the float-free release of the survival craft from its stowed position shall be automatic.

6.1.4 *Launching appliances for free-fall lifeboats*

6.1.4.1 Every free-fall launching appliance shall comply with the applicable requirements of paragraph 6.1.1 and, in addition, shall comply with the requirements of this paragraph.

6.1.4.2 The launching appliance shall be designed and installed so that it and the lifeboat it serves operate as a system to protect the occupants from harmful acceleration forces as required by paragraph 4.7.5, and to ensure effective clearing of the ship as required by paragraphs 4.7.3.1 and 4.7.3.2.

6.1.4.3 The launching appliance shall be constructed so as to prevent sparking and incendiary friction during the launching of the lifeboat.

6.1.4.4 The launching appliance shall be designed and arranged so that, in its ready-to-launch position, the distance from the lowest point on the lifeboat it serves to the water surface with the ship in its lightest seagoing condition does not exceed the lifeboat's free-fall certification height, taking into consideration the requirements of paragraph 4.7.3.

6.1.4.5 The launching appliance shall be arranged so as to preclude accidental release of the lifeboat in its unattended stowed position. If the means provided to secure the lifeboat cannot be released from inside the lifeboat, it shall be so arranged as to preclude boarding the lifeboat without first releasing it.

6.1.4.6 The release mechanism shall be arranged so that at least two independent actions from inside the lifeboat are required in order to launch the lifeboat.

6.1.4.7 Each launching appliance shall be provided with a secondary means to launch the lifeboat by falls. Such means shall comply with the requirements of paragraph 6.1.1 (except 6.1.1.3) and paragraph 6.1.2 (except 6.1.2.6). It must be capable of launching the lifeboat against unfavourable conditions of trim of up to only 2° and list of up to only 5° either way and it need not comply with the speed requirements of paragraphs 6.1.2.8 and 6.1.2.9. If the secondary launching appliance is not dependent on gravity, stored mechanical power or other manual means, the launching appliance shall be connected both to the ship's main and emergency power supplies.

6.1.4.8 The secondary means of launching shall be equipped with at least a single off-load capability to release the lifeboat.

6.1.5 *Liferaft launching appliances*

Every liferaft launching appliance shall comply with the requirements of paragraphs 6.1.1 and 6.1.2, except with regard to embarkation in the stowed position, recovery of the loaded liferaft and that manual operation is permitted for turning out the appliance. The launching appliance shall include an automatic release hook arranged so as to prevent premature release during lowering and shall release the liferaft when waterborne. The release hook shall include a capability to release the hook under load. The on-load release control shall:

- **.1** be clearly differentiated from the control which activates the automatic release function;
- **.2** require at least two separate actions to operate;
- **.3** with a load of 150 kg on the hook, require a force of at least 600 N and not more than 700 N to release the load, or provide equivalent adequate protection against inadvertent release of the hook; and
- **.4** be designed such that the crew members on deck can clearly observe when the release mechanism is properly and completely set.

6.1.6 *Embarkation ladders*

6.1.6.1 Handholds shall be provided to ensure a safe passage from the deck to the head of the ladder and vice versa.

6.1.6.2 The steps of the ladder shall be:

- **.1** made of hardwood, free from knots or other irregularities, smoothly machined and free from sharp edges and splinters, or of suitable material of equivalent properties;
- **.2** provided with an efficient non-slip surface either by longitudinal grooving or by the application of an approved non-slip coating;
- **.3** not less than 480 mm long, 115 mm wide and 25 mm in depth, excluding any non-slip surface or coating; and
- **.4** equally spaced not less than 300 mm or more than 380 mm apart and secured in such a manner that they will remain horizontal.

6.1.6.3 The side ropes of the ladder shall consist of two uncovered manila ropes not less than 65 mm in circumference on each side. Each rope shall be continuous with no joints below the top step. Other materials may be used provided the dimensions, breaking strain, weathering, stretching and gripping properties are at least equivalent to those of manila rope. All rope ends shall be secured to prevent unravelling.

6.1.7 *Launching appliances for fast rescue boats*

6.1.7.1 Every fast rescue boat launching appliance shall comply with the requirements of paragraphs 6.1.1 and 6.1.2 except 6.1.2.10 and, in addition, shall comply with the requirements of this paragraph.

6.1.7.2 The launching appliance shall be fitted with a device to dampen the forces due to interaction with the waves when the fast rescue boat is launched or recovered. The device shall include a flexible element to soften shock forces and a damping element to minimize oscillations.

6.1.7.3 The winch shall be fitted with an automatic high-speed tensioning device which prevents the wire from going slack in all sea state conditions in which the fast rescue boat is intended to operate.

6.1.7.4 The winch brake shall have a gradual action. When the fast rescue boat is lowered at full speed and the brake is applied sharply, the additional dynamic force induced in the wire due to retardation shall not exceed 0.5 times the working load of the launching appliance.

6.1.7.5 The lowering speed for a fast rescue boat with its full complement of persons and equipment shall not exceed 1 m/s. Notwithstanding the requirements of paragraph 6.1.1.9, a fast rescue boat launching appliance shall be capable of hoisting the fast rescue boat with 6 persons and its full complement of equipment at a speed of not less than 0.8 m/s. The appliance shall also be capable of lifting the rescue boat with the maximum number of persons that can be accommodated in it, as calculated in accordance with paragraph 4.4.2.

6.2 Marine evacuation systems

6.2.1 *Construction of the marine evacuation systems*

6.2.1.1 The passage of the marine evacuation system shall provide for safe descent of persons of various ages, sizes and physical capabilities, wearing approved lifejackets, from the embarkation station to the floating platform or survival craft.

6.2.1.2 Strength and construction of the passage and platform shall be to the satisfaction of the Administration.

6.2.1.3 The platform, if fitted, shall be:

.1 such that sufficient buoyancy will be provided for the working load. In the case of an inflatable platform, the main buoyancy chambers, which for this purpose shall include any thwarts or floor inflatable structural members, are to meet the requirements of section 4.2 based upon the platform capacity, except that the capacity shall be obtained by dividing by 0.25 the usable area given in paragraph 6.2.1.3.3;

.2 stable in a seaway and shall provide a safe working area for the system operators;

.3 of sufficient area that will provide for the securing of at least two liferafts for boarding and to accommodate at least the number of persons that at any time are expected to be on the platform. This usable platform area shall be at least equal to:

$$\frac{\text{20\% of total number of persons that the marine evacuation system is certified for}}{4}\ \text{m}^2$$

or 10 m^2, whichever is the greater. However, Administrations may approve alternate arrangements which are demonstrated to comply with all of the prescribed performance requirements;*

* Refer to the Revised recommendation on testing of life saving appliances adopted by the Organization by resolution MSC.81(70), as amended.

.4 self-draining;

.5 subdivided in such a \
compartment will not
of evacuation. The b
protected against da
ship's side;

.6 fitted with a stabil
Administration;

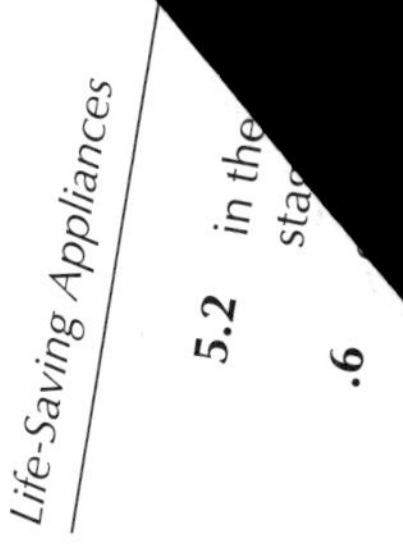

.7 restrained by a b
which are designed to deplo
to be capable of being adjusted to the pos
evacuation; and

.8 provided with mooring and bowsing line patches of sufficient strength to securely attach the largest inflatable liferaft associated with the system.

6.2.1.4 If the passage gives direct access to the survival craft, it should be provided with a quick-release arrangement.

6.2.2 *Performance of the marine evacuation system*

6.2.2.1 A marine evacuation system shall be:

.1 capable of deployment by one person;

.2 such as to enable the total number of persons for which it is designed, to be transferred from the ship into the inflated liferafts within a period of 30 min in the case of a passenger ship and of 10 min in the case of a cargo ship from the time the abandon ship signal is given;

.3 arranged such that liferafts may be securely attached to the platform and released from the platform by a person either in the liferaft or on the platform;

.4 capable of being deployed from the ship under unfavourable conditions of trim of up to 10° and list of up to 20° either way;

.5 in the case of being fitted with an inclined slide, such that the angle of the slide to the horizontal is:

5.1 within a range of 30° to 35° when the ship is upright and in the lightest sea-going condition; and [illegible] case of a passenger ship, a maximum of 55° in the final [illegible]ge of flooding set by the requirements in regulation II-1/8;

[illegible] evaluated for capacity by means of timed evacuation deployments conducted in harbour;

.7 capable of providing a satisfactory means of evacuation in a sea state associated with a wind of force 6 on the Beaufort scale;

.8 designed to, as far as practicable, remain effective under conditions of icing; and

.9 so constructed that only a minimum amount of routine maintenance is necessary. Any part requiring maintenance by the ship's crews shall be readily accessible and easily maintained.

6.2.2.2 Where one or more marine evacuation systems are provided on a ship, at least 50% of such systems shall be subjected to a trial deployment after installation. Subject to these deployments being satisfactory, the untried systems are to be deployed within 12 months of installation.

6.2.3 *Inflatable liferafts associated with marine evacuation systems*

Any inflatable liferaft used in conjunction with the marine evacuation system shall:

.1 conform with the requirements of section 4.2;

.2 be sited close to the system container but be capable of dropping clear of the deployed system and boarding platform;

.3 be capable of release one at a time from its stowage rack with arrangements which will enable it to be moored alongside the platform;

.4 be stowed in accordance with regulation III/13.4; and

.5 be provided with pre-connected or easily connected retrieving lines to the platform.

6.2.4 *Containers for marine evacuation systems*

6.2.4.1 The evacuation passage and platform shall be packed in a container that is:

.1 so constructed as to withstand hard wear under conditions encountered at sea; and

.2 as far as practicable watertight, except for drain holes in the container bottom.

6.2.4.2 The container shall be marked with:

- **.1** maker's name or trademark;
- **.2** serial number;
- **.3** name of approval authority and the capacity of the system;
- **.4** SOLAS;
- **.5** date of manufacture (month and year);
- **.6** date and place of last service;
- **.7** maximum permitted height of stowage above waterline; and
- **.8** stowage position on board.

6.2.4.3 Launching and operating instructions shall be marked on or in the vicinity of the container.

6.2.5 *Marking on marine evacuation systems*

The marine evacuation system shall be marked with:

- **.1** maker's name or trademark;
- **.2** serial number;
- **.3** date of manufacture (month and year);
- **.4** name of approving authority;
- **.5** name and place of servicing station where it was last serviced, along with the date of servicing; and
- **.6** the capacity of the system.

Chapter VII
Other life-saving appliances

7.1 Line-throwing appliances

7.1.1 Every line-throwing appliance shall:

- **.1** be capable of throwing a line with reasonable accuracy;
- **.2** include not less than four projectiles, each capable of carrying the line at least 230 m in calm weather;
- **.3** include not less than four lines, each having a breaking strength of not less than 2 kN; and
- **.4** have brief instructions or diagrams clearly illustrating the use of the line-throwing appliance.

7.1.2 The rocket, in the case of a pistol-fired rocket, or the assembly, in the case of an integral rocket and line, shall be contained in a water-resistant casing. In addition, in the case of a pistol-fired rocket, the line and rockets together with the means of ignition shall be stowed in a container which provides protection from the weather.

7.2 General alarm and public address system

7.2.1 *General emergency alarm system*

7.2.1.1 The general emergency alarm system shall be capable of sounding the general emergency alarm signal consisting of seven or more short blasts followed by one long blast on the ship's whistle or siren and additionally on an electrically operated bell or klaxon or other equivalent warning system, which shall be powered from the ship's main supply and the emergency source of electrical power required by regulation II-1/42 or II-1/43, as appropriate. The system shall be capable of operation from the navigation bridge and, except for the ship's whistle, also from other strategic points. The alarm shall continue to function after it has been triggered until it is manually turned off or is temporarily interrupted by a message on the public address system.

7.2.1.2 The minimum sound pressure levels for the emergency alarm tone in interior and exterior spaces shall be 80 dB (A) and at least 10 dB (A) above ambient noise levels existing during normal equipment operation with the ship under way in moderate weather.

7.2.1.3 The sound pressure levels at the sleeping position in cabins and in cabin bathrooms shall be at least 75 dB (A) and at least 10 dB (A) above ambient noise levels.*

7.2.2 *Public address system*

7.2.2.1 The public address system shall be a loudspeaker installation enabling the broadcast of messages into all spaces where crew members or passengers, or both, are normally present, and to muster stations. It shall allow for the broadcast of messages from the navigation bridge and such other places on board the ship as the Administration deems necessary. It shall be installed with regard to acoustically marginal conditions and not

* Refer to the *Code on Alerts and Indicators, 2009*, adopted by the Organization by resolution A.1021(26).

require any action from the addressee. It shall be protected against unauthorized use.

7.2.2.2 With the ship under way in normal conditions, the minimum sound pressure levels for broadcasting emergency announcements shall be:

.1 in interior spaces 75 dB (A) and at least 20 dB (A) above the speech interference level; and

.2 in exterior spaces 80 dB (A) and at least 15 dB (A) above the speech interference level.

Testing and Evaluation of Life-Saving Appliances

I
Revised recommendation on testing of life-saving appliances

Resolution MSC.81(70)

(adopted on 11 December 1998)

THE MARITIME SAFETY COMMITTEE,

RECALLING Article 28(b) of the Convention on the International Maritime Organization concerning the functions of the Committee,

RECALLING ALSO that the Assembly, when adopting resolution A.689(17) on Testing of life-saving appliances, authorized the Committee to keep the Recommendation on testing of life-saving appliances under review and to adopt, when appropriate, amendments thereto,

RECALLING FURTHER that, since the adoption of resolution A.689(17), the Committee has amended the Recommendation annexed thereto in five occasions, i.e., by resolution MSC.54(66), by circulars MSC/Circ.596, MSC/Circ.615 and MSC/Circ.809 and through the present resolution,

NOTING that the 1995 SOLAS Conference, in adopting amendments to the 1974 SOLAS Convention concerning the safety of ro-ro passenger ships, also adopted resolution 7 on Development of requirements, guidelines and performance standards, whereby the Committee was requested to develop relevant requirements, guidelines and performance standards to assist in the implementation of the amendments adopted by the Conference,

NOTING ALSO that the requirements of the International Life-Saving Appliance (LSA) Code came into force on 1 July 1998 under the provisions of new chapter III of the 1974 SOLAS Convention,

RECOGNIZING the need to introduce more precise requirements in the testing of life-saving appliances,

HAVING CONSIDERED the recommendation made by the Sub-Committee on Ship Design and Equipment at its forty-first session,

1. ADOPTS the Revised Recommendation on Testing of Life-Saving Appliances, set out in the annex to the present resolution;

2. RECOMMENDS Governments to ensure that life-saving appliances are subjected to:

.1 the tests recommended in the annex to the present resolution; or

.2 such tests as the Administration is satisfied are substantially equivalent to those recommended in the annex to the present resolution.

Introduction

The tests in this Recommendation have been developed on the basis of the requirements of the International Life-Saving Appliances (LSA) Code.

Life-saving appliances which are installed on board on or after 1 July 2010 should meet the applicable requirements of this Recommendation or substantially equivalent ones, as may be specified by the Administration. Where there has been a substantial change in the equipment performance requirements or the test procedures in this recommendation, an item of equipment previously tested to resolution A.521(13), or previous versions of resolution A.689(17), need only be subjected to tests affected by such changes.

Life-saving appliances which were installed on board before 1 July 2010 may meet the applicable requirements of the Recommendation on Testing of Life-Saving Appliances adopted by resolution A.521(13), previous versions of resolution A.689(17), or substantially equivalent ones, as may be specified by the Administration, and may continue in use on the ship on which they are presently installed, as long as they remain suitable for service.

Tests for requirements referred to in the LSA Code, which are not included in this Recommendation, should be to the satisfaction of the Administration.

It should be verified that life-saving appliances not covered by tests referred to in this Recommendation meet the applicable requirements of the LSA Code.

Part 1
Prototype tests for life-saving appliances

1 Lifebuoys

1.1 Lifebuoys specification

It should be established by measurement, weighing and inspection that:

.1 the lifebuoy has an outer diameter of not more than 800 mm and an inner diameter of not less than 400 mm;

.2 the lifebuoy has a mass of not less than 2.5 kg;

.3 if it is intended to operate the quick-release arrangement provided for a self-activated smoke signal and self-igniting light, the lifebuoy has a mass of not less than 4 kg (see 1.8); and

.4 the lifebuoy is fitted with a grabline of not less than 9.5 mm in diameter and of not less than four times the outside diameter of the body of the buoy in length and secured in four equal loops.

1.2 Temperature cycling test

The following test should be carried out on two lifebuoys.

1.2.1 The lifebuoys should be alternately subjected to surrounding temperatures of –30°C and +65°C. These alternating cycles need not follow immediately after each other and the following procedure, repeated for a total of 10 cycles, is acceptable:

.1 an 8 h exposure at a minimum temperature of +65°C to be completed in one day; and

.2 the specimens removed from the warm chamber that same day and left exposed under ordinary room conditions at a temperature of 20°C ± 3°C until the next day;

.3 an 8 h exposure at a maximum temperature of –30°C to be completed the next day; and

.4 the specimens removed from the cold chamber that same day and left exposed under ordinary room conditions at a temperature of 20°C ± 3°C until the next day.

1.2.2 The lifebuoys should show no sign of loss of rigidity under high temperatures and, after the tests, should show no sign of damage such as shrinking, cracking, swelling, dissolution or change of mechanical qualities.

1.3 Drop test

Each lifebuoy should be suspended from its upper edge via a release device so that the lower edge of the lifebuoy is at the height at which it is intended to be stowed on ships in their lightest seagoing condition, or 30 m, whichever is the greater, and dropped into the water without suffering damage. In addition, one lifebuoy should be suspended from its upper edge via a release device so that the lower edge of the lifebuoy is at a height of 2 m, and dropped three times onto a concrete floor, without suffering damage.

1.4 Test for oil resistance

One of the lifebuoys should be immersed horizontally for a period of 24 h under a 100 mm head of diesel oil at normal room temperature. After this test the lifebuoy should show no sign of damage such as shrinking, cracking, swelling, dissolution or change of mechanical qualities.

1.5 Fire test

The other lifebuoy should be subjected to a fire test. A test pan, 30 cm × 35 cm × 6 cm, should be placed in an essentially draught-free area. Water should be put into the bottom of the test pan to a depth of 1 cm followed by enough petrol to make a minimum total depth of 4 cm. The petrol should then be ignited and allowed to burn freely for 30 s. The lifebuoy should then be moved through flames in an upright, forward, free-hanging position, with the bottom of the lifebuoy 25 cm above the top edge of the test pan so that the duration of exposure to the flames is 2 s. The lifebuoy should not sustain burning or continue melting after being removed from the flames.

1.6 Flotation test

The two lifebuoys subjected to the above tests should be floated in fresh water with not less than 14.5 kg of iron suspended from each of them and should remain floating for a period of 24 h.

1.7 Strength test

A lifebuoy body should be suspended by a 50 mm wide strap. A similar strap should be passed around the opposite side of the body with a 90 kg mass suspended from it. After 30 min, the lifebuoy body should be examined. There should be no breaks, cracks or permanent deformation.

1.8 Test for operation with a light and smoke signal

A lifebuoy intended for quick release with a light and smoke signal should be given this test. The lifebuoy should be arranged in a manner simulating its installation on a ship for release from the navigating bridge. A lifebuoy light and smoke signal should be attached to the lifebuoy in the manner recommended by the manufacturers. The lifebuoy should be released and should activate both the light and the smoke signal.

1.9 Lifebuoy self-activating smoke signal tests

1.9.1 Nine self-activating smoke signals should be subjected to temperature cycling as prescribed in 1.2.1 and, after the tests, should show no sign of damage such as shrinking, cracking, swelling, dissolution or change of mechanical qualities.

1.9.2 After at least 10 complete temperature cycles, the first three smoke signals should be subjected to a temperature of –30°C for at least 48 h, then taken from this stowage temperature and be activated and operated in seawater at a temperature of –1°C and the next three smoke signals should be subjected to a temperature of +65°C for at least 48 h then taken from this stowage temperature and be activated and operated in seawater at a temperature of +30°C. After the smoke signals have been emitting smoke for 7 min, the smoke-emitting ends of the smoke signals should be immersed to a depth of 25 mm for 10 s. On being released the smoke signals should continue operating for a total period of smoke emission of not less than 15 min. The signals should not ignite explosively or in a manner dangerous to persons close by.

1.9.3 The last three smoke signals taken from ordinary room conditions and attached by a line to a lifebuoy having a mass of not more than 4 kg should undergo the drop test into water prescribed in 1.3. The lifebuoy should have both a smoke signal and a lifebuoy light attached in the manner recommended by the manufacturers and be dropped from a quick-release fitting. The smoke signals should not be damaged and should function for a period of at least 15 min.

1.9.4 Smoke signals should also be subjected to the tests and examinations prescribed in 4.2.4, 4.3.1, 4.3.3, 4.5.5, 4.5.6, 4.8.2 and 4.8.3.

1.9.5 A smoke signal should be tested in waves at least 300 mm high. The signal should function effectively and for not less than 15 min.

1.9.6 A force of 225 N should be applied to the fitting that attaches the self-activating smoke signal to the lifebuoy. Neither the fitting nor the signal should be damaged as a result of the test.

2 Lifejackets

2.1 Temperature cycling test

A lifejacket should be subjected to the temperature cycling as prescribed in 1.2.1 and should then be externally examined. The lifejacket materials should show no sign of damage such as shrinking, cracking, swelling, dissolution or change of mechanical qualities.

2.2 Buoyancy test

The buoyancy of the lifejacket should be measured before and after 24 h complete submersion to just below the surface in fresh water. The difference between the initial buoyancy and the final buoyancy should not exceed 5% of the initial buoyancy.

2.3 Fire test

A lifejacket should be subjected to the fire test prescribed in 1.5. The lifejacket should not sustain burning for more than 6 s or continue melting after being removed from the flames.

2.4 Tests of components other than buoyancy materials

All the materials, other than buoyancy materials, used in the construction of the lifejacket, including the cover, tapes, seams and closures should be tested to an international standard acceptable to the Organization* to establish that they are rot-proof, colour fast and resistant to deterioration from exposure to sunlight and that they are not unduly affected by seawater, oil or fungal attack.

2.5 Strength tests

Body or lifting loop strength tests

2.5.1 The lifejacket should be immersed in water for a period of 2 min. It should then be removed from the water and closed in the same manner as when it is worn by a person. A force of not less than 3,200 N (2,400 N in the case of a child or infant-size lifejacket) should be applied for 30 min to the

* Refer to the recommendations of the International Organization for Standardization, in particular publication ISO 12402-7, *Personal flotation devices – Part 7: Materials and components – Safety requirements and test methods*.

part of the lifejacket that secures it to the body of the wearer (see figure 1) and separately to the lifting loop of the lifejacket. The lifejacket should not be damaged as a result of this test. The test should be repeated for each encircling closure.

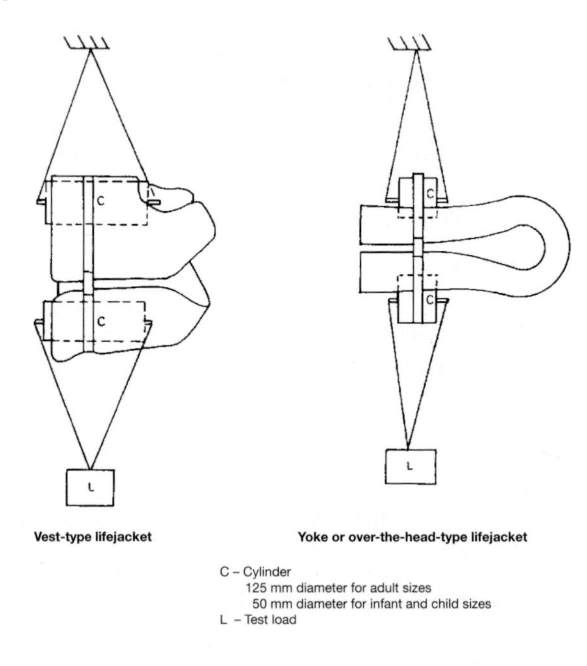

Figure 1 – *Body strength test arrangement for lifejackets*

Shoulder lift test

2.5.2 The lifejacket should be immersed in water for a period of 2 min. It should then be removed from the water and closed on a form as shown in figure 2, in the same manner as when it is worn by a person. A force of not less than 900 N (700 N in the case of a child- or infant-size lifejacket) should be applied for 30 min across the form and the shoulder section of the lifejacket (see figure 3). The lifejacket should not be damaged as a result of this test. The lifejacket should remain secured on the form during this test.

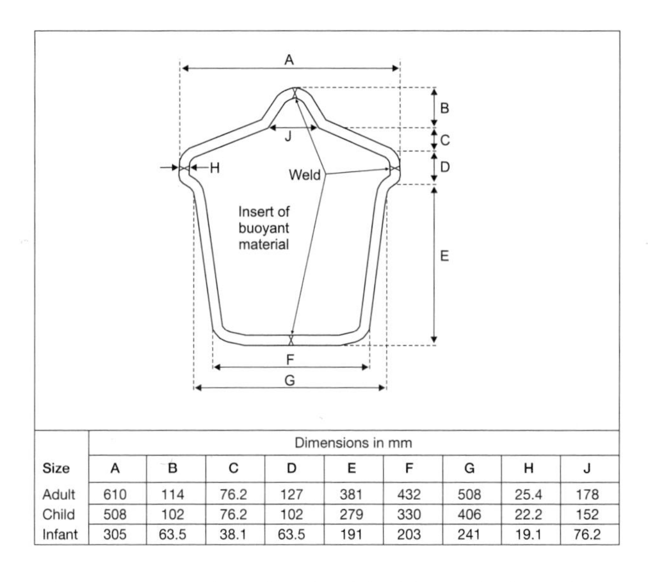

Size	Dimensions in mm								
	A	B	C	D	E	F	G	H	J
Adult	610	114	76.2	127	381	432	508	25.4	178
Child	508	102	76.2	102	279	330	406	22.2	152
Infant	305	63.5	38.1	63.5	191	203	241	19.1	76.2

Figure 2 – *Test form for shoulder lift test for lifejackets*

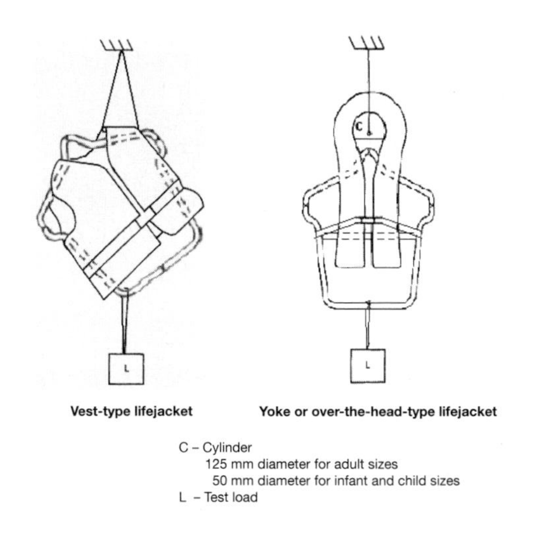

Figure 3 – *Shoulder lift test arrangement for lifejackets*

2.6 Tests for lifejacket buoyancy material

The following tests should be carried out on eight specimens of each lifejacket buoyancy material. The specimens should be at least 300 mm square and be of the same thickness as used in the lifejacket. In the case of kapok, the entire lifejacket should be subjected to the test. The dimensions should be recorded at the beginning and end of these tests. Where multiple layers of materials are used to achieve the total thickness desired for the lifejacket, the specimens should be of the thinnest material used.

Test for stability under temperature cycling

2.6.1 Six specimens should be subjected to temperature cycling as prescribed in 1.2.1.

2.6.2 The dimensions of the specimens (except kapok) should be recorded at the end of the last cycle. The specimens should be carefully examined and should not show any sign of external change of structure or of mechanical qualities.

2.6.3 Two of the specimens should be cut open and should not show any sign of internal change of structure.

2.6.4 Four of the specimens should be used for compression and water absorption tests, two of which should be so tested after they have also been subjected to the diesel oil test as prescribed in 1.4.

Tests for compression and water absorption

2.6.5 The tests should be carried out in fresh water and the specimens should be immersed for a period of seven days under a 1.25 m head of water.

2.6.6 The tests should be carried out:

.1 on two specimens as supplied;

.2 on two specimens which have been subjected to the temperature cycling as prescribed in 2.6.1; and

.3 on two specimens which have been subjected to the temperature cycling as prescribed in 2.6.1 followed by the diesel oil test as prescribed in 1.4.

2.6.7 The results should state the buoyant force in N which each specimen exerts when submerged in water after one and seven days' immersion. The reduction of buoyancy should not exceed 10% for specimens which have been exposed to the diesel oil conditioning and must not exceed 5% for all other specimens. The specimens should show no sign of damage such as shrinking, cracking, swelling, dissolution or change of mechanical qualities.

Tensile strength test

2.6.8 The tensile strength at break of the material should be measured before and after the combined exposure described in 2.6.6.3. When tested according to an international standard acceptable to the Organization,* the materials should have a minimum tensile strength of 140 kPa before exposure, which should not be reduced by more than 25% following the combined exposures. In the case of kapok, the protective cover should have a minimum breaking strength of 13 kPa before exposure, which should not be reduced by more than 25% following the combined exposures.

2.7 Donning test

2.7.1 To minimize the risk of incorrect donning by uninitiated persons, often in adverse conditions, lifejackets should be examined for the following features and tested as follows:

- **.1** fastenings necessary for proper performance should be few and simple, and provide quick and positive closure that does not require tying of knots;
- **.2** adult lifejackets should readily fit various sizes of adults, both lightly and heavily clad; and
- **.3** all lifejackets should be capable of being worn inside-out, or clearly in only one way.

Test subjects

2.7.2 These tests should be carried out with at least 12 able-bodied persons who are completely unfamiliar with the lifejacket and selected according to the heights and weights in table 2.1 and the following:

- **.1** small test subjects need not be adults;
- **.2** at least 1/3, but not more than 1/2 of test subjects should be females, including at least 1 per height category but excluding the tallest height;

* Refer to the recommendations of the International Organization for Standardization, in particular publication ISO 12402-7, *Personal flotation devices – Part 7: Materials and components – Safety requirements and test methods.*

.3 at least one male and one female should be from the lowest and highest weight group;

.4 at least one subject should be selected from each cell containing a "1"; and

.5 enough additional subjects should be selected from cells containing a "X" to total the required number of test subjects, with no more than one subject per cell. A uniform distribution across weight ranges should be maintained.

Table 2.1 – *Test subject selection for adult lifejackets*

Height range (m)	Weight range – kg							
	40–43	43–60	60–70	70–80	80–100	100–110	110–120	> 120
< 1.5	1	X	X	X				
1.5–1.6	X	1	1	X	X			
1.6–1.7		X	X	1	X	X		
1.7–1.8			X	X	1	X	X	X
1.8–1.9			X	X	X	1	1	X
> 1.9					X	X	X	1

Clothing

2.7.3 Each test subject should be tested wearing the clothing specified for the test and appropriate to their size as follows:

.1 *Normal clothing* means normal indoor clothing, which would not normally interfere with the donning of a lifejacket; and

.2 *Heavy-weather clothing* means the attire appropriate for a hostile environment, including a hooded arctic parka and warm cotton gloves.

2.7.4 Each test should be timed from when the order is given until the test subject declares that donning is complete. For assessment purposes donning is considered complete when the subject has donned and securely adjusted all methods of securing the lifejacket to the extent needed to meet the in-water performance requirements, including inflation, if needed.

Test without instruction

2.7.4.1 The test subjects may be tested individually or as a group. Wearing normal clothing, the first attempt should be with no assistance, guidance or prior demonstration. The lifejacket, with closures in the stored condition, should be placed on the floor, face up, in front of the test subject. The instruction provided should be identical for each subject and should be equivalent to the following: "PLEASE DON THIS LIFEJACKET AS QUICKLY AS POSSIBLE AND ADJUST IT TO A SNUG FIT SO YOU CAN ABANDON SHIP." The lifejacket should be capable of being donned by at least 75% of the subjects, and within 1 min. If a subject dons the lifejacket substantially correctly but fails to secure and/or adjust all closures, the jump test in 2.8.8 and in-water performance tests in 2.8.5 and 2.8.6 should be performed with the lifejacket as donned to establish whether the performance is acceptable and the donning is successful.

Test after instruction

2.7.4.2 For each subject whose first attempt exceeds 1 min or is incomplete, after demonstration or instruction to familiarize the subject with the donning procedure, the test subject should then don the lifejacket without assistance while wearing normal clothing, using the same instruction and timing method as in 2.7.4.1. Each subject should correctly don the lifejacket within a period of 1 min.

Heavy-weather clothing test

2.7.4.3 Each subject should then don the lifejacket without assistance while wearing heavy weather clothing, using the same instruction and timing method as in 2.7.4.1. Each subject should don the lifejacket correctly within a period of 1 min.

2.8 Water performance tests

2.8.1 This portion of the test is intended to determine the ability of the lifejacket to assist a helpless person or one in an exhausted or unconscious state and to show that the lifejacket does not unduly restrict movement. The in-water performance of a lifejacket is evaluated by comparison to the performance of a suitable size standard reference lifejacket, i.e., Reference Test Device (RTD) as specified in annexes 1 to 3. All tests should be carried out in fresh water under still conditions.

Test subjects

2.8.2 These tests should be carried out with at least 12 persons as described in 2.7.2. Only good swimmers should be used, since the ability to relax in the water is rarely otherwise obtained.

Clothing

2.8.3 Subjects should wear only swimming costumes.

Preparation for water performance tests

2.8.4 The test subjects should be made familiar with each of the tests set out below, particularly the requirement regarding relaxing and exhaling in the face-down position. The test subject should don the lifejacket, unassisted, using only the instructions provided by the manufacturer. After entering the water, care should be taken to ensure that there is no significant amount of air unintentionally trapped in the lifejacket or swimming costume.

Righting tests

2.8.5 Each test subject should assume a prone, face down position in the water, but with the head lifted up so the mouth is out of the water. The subject's feet should be supported, shoulder width apart, with the heels just below the surface of the water. After assuming a starting position with the legs straight and arms along the sides, the subject should then be instructed in the following sequence to allow the body to gradually and completely relax into a natural floating posture: allow the arms and shoulders to relax; allow the legs to relax; and then the spine and neck, letting the head fall into the water while breathing out normally. During the relaxation phase, the subject should be maintained in a stable face down position. Immediately after the subject has relaxed with the face in the water, simulating a state of utter exhaustion, the subject's feet should be released. The period of time until the mouth of the test subject comes clear of the water should be recorded to the nearest 1/10 of a second, starting from when the subject's feet are released. The above test should be conducted for a total of six times, and the highest and lowest times discarded. The test should then be conducted for a total of six times in the RTD and the highest and lowest times discarded.

Static balance measurements

2.8.6 At the conclusion of the righting tests without making any adjustments in body or lifejacket position, measurements should be made with the subject floating in the relaxed face-up position of static balance resulting from the preceding tests. The following measurements should be made (see figure 4):

.1 freeboard – the distance measured perpendicularly from the surface of the water to the lowest point of the subject's mouth where respiration may be impeded, if the mouth were not held shut. The lowest side of the mouth should be measured if the left and right sides are not level;

.2 faceplane angle – the angle, relative to the surface of the water, of the plane formed between the most forward part of the forehead and the chin;

.3 torso angle – the angle, relative to vertical, of the line formed by the forward points of the shoulder and hipbone (ilium portion of the pelvis); and

.4 list angle – the angle relative to the surface of the water and a line between the left and right shoulder or a line through the ears if only the head is tilted.

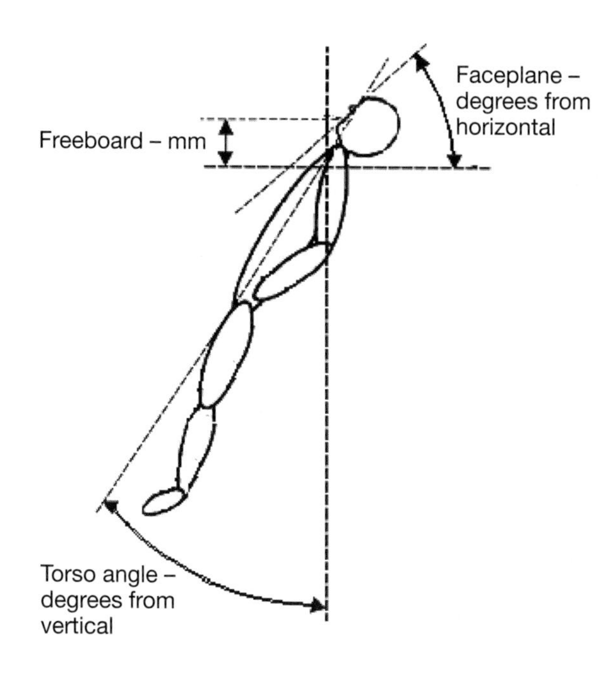

Figure 4 – *Static balance measurements*

Assessment

2.8.7 After the water tests described in 2.8.5 and .6 above:

.1 *Turning time:* The average turn time for all subjects in the candidate lifejacket should not exceed the average time in the RTD, and the number of "no-turns", if any, should not exceed the number in the RTD;

.2 *Freeboard:* The average freeboard of all the subjects should not be less than the average for the RTD;

.3 *Torso angles:* The average of all subjects' torso angles should be not less than the average for the RTD minus 5°;

.4 *Faceplane (head) angles:* The average of all subjects' faceplane angles should be not less than the average for the RTD minus 5°;

.5 *Lifejacket light location:* The position of the lifejacket light should permit it to be visible over as great a segment of the upper hemisphere as is practicable.

Jump and drop tests

2.8.8 Without readjusting the lifejacket, the test subject should jump vertically into the water, feet first, from a height of at least 1 m while holding the arms over the head. Upon entering the water, the test subject should relax to simulate a state of utter exhaustion. The freeboard to the mouth should be recorded after the test subject comes to rest. The test should be repeated from a height of at least 4.5 m but, when jumping into the water, the test subject should hold onto the lifejacket during water entry to avoid possible injury. Upon entering the water, the test subject should relax to simulate a state of utter exhaustion. The freeboard to the mouth should be recorded after the test subject comes to rest. The lifejacket and its attachments should be examined for any damage. If injury is believed likely from any jump or drop test the lifejacket should be rejected or the test delayed until tests from a lower height or with additional precautions demonstrate that the risk from the required test is acceptable.

Assessment

2.8.9 Following the drop test, the lifejacket should:

.1 surface the test subject in a face-up position with an average freeboard for all the subjects of not less than the average for the RTD determined in accordance with 2.8.6;

.2 not be dislodged or cause harm to the test subject;

.3 have no damage that would affect its in-water performance or buoyancy; and

.4 have no damage to its attachments.

Stability test

2.8.10 The test subject should attain a relaxed face-up position of static balance in the water. The subject should be instructed to assume a foetal position as follows: "place your elbows against your sides, your hands on your stomach, under the lifejacket if possible, and bring your knees up as close to your chest as possible." The subject should be rotated clockwise around the longitudinal axis of the torso by grasping the subject's shoulders or upper areas of the lifejacket so that the subject attains a 55 ± 5 degree list. The subject should then be released. The subject should return to a stable face-up position. The test should then be conducted with the subject rotated counter-clockwise. The entire test should then be repeated with the test subject wearing the RTD. The candidate lifejacket should not roll any subject face down in the water. The number of subjects who are returned to the stable face-up foetal position in the candidate lifejacket should be at least equal to the number who are returned to the stable face-up foetal position in the RTD.

Swimming and water emergence test

2.8.11 All test subjects, without wearing the lifejacket, should attempt to swim 25 m and board a liferaft or a rigid platform with its surface 300 mm above the water surface. All test subjects who successfully complete this task should perform it again wearing the lifejacket. At least two thirds of the test subjects who can accomplish the task without the lifejacket should also be able to perform it with the lifejacket.

2.9 Infant and children's lifejacket tests

As far as possible, similar tests should be applied for approval of lifejackets suitable for infants and children.

Infant and child test subjects

2.9.1 For child-size lifejackets, tests should be carried out with at least 9 able-bodied persons, and for infant-size lifejackets, tests should be carried out with at least 5 able-bodied persons. All test subjects should be selected according to table 2.2 or table 2.3 as follows:

.1 One subject should be selected per each cell containing a "1".

.2 Remaining subjects should be selected from cells containing an "X", without repeating a cell.

.3 At least 40% of the subjects should be male and at least 40% female.

.4 Devices for infants should be tested on infants as small as 6 kg mass.

.5 A manikin or manikins may be substituted for infant lifejacket test subjects if the manikin or manikins have been demonstrated to provide representative results compared to human subjects.

Table 2.2 – *Selection of child test subjects*

Height range (cm)	**Weight range (kg)**										
	14–17	**17–20**	**20–22**	**22–25**	**25–28**	**28–30**	**30–33**	**33–36**	**36–38**	**38–41**	**41–43**
79–105	1	X									
90–118		X	1								
102–130				1	X						
112–135					X	1					
122–150							1	1	X		
145–165									X	1	1

Table 2.3 – *Selection of infant test subjects*

Height range (cm)	**Weight range (kg)**		
	Less than 11	**11–14**	**14–17**
Less than 83	1	X	
79–105	X	1	1
90–118			X

2.9.2 When conducting water performance tests under 2.8, infant and child-size lifejackets should meet the following requirements for their critical flotation stability characteristics:

- **.1** *Turning time:* The average turn time for all subjects in the candidate lifejacket should not exceed the average time in the appropriate size RTD;
- **.2** *Freeboard:* The average results for clearance of the mouth above the water for all subjects should not be less than the average for the appropriate size RTD;
- **.3** *Torso angle:* The average of all subjects' results should be not less than the average for the appropriate size RTD minus 10°;
- **.4** *Faceplane (head) angle:* The average of all subjects' results should be not less than the average for the appropriate size RTD minus 10°; and
- **.5** *Mobility:* Mobility of the subject both in and out of the water should be given consideration in determining the acceptability of a device for approval and should be compared to mobility when wearing the appropriate size RTD when climbing out of the water, going up and down stairs, picking up an article from the floor, and then drinking from a cup.

2.9.3 With the exception of reducing freeboard and self-righting performance, the requirements for infant lifejackets may be relaxed if necessary in order to:

- **.1** contribute to the rescue of the infant by a caretaker;
- **.2** allow the infant to be fastened to a caretaker and contribute to keeping the infant close to the caretaker;
- **.3** keep the infant dry, with free respiratory passages;
- **.4** protect the infant against bumps and jolts during the evacuation; and
- **.5** allow a caretaker to monitor and control heat loss by the infant.

2.10 Tests for inflatable lifejackets

2.10.1 *Inflation tests*

2.10.1.1 Two inflatable lifejackets should be subjected to the temperature cycling test prescribed in 1.2.1 in the uninflated condition and should then be externally examined. The inflatable lifejacket materials should show no sign of damage such as shrinking, cracking, swelling, dissolution or change of mechanical qualities. The automatic and manual inflation systems should each be tested immediately after each temperature cycling test as follows:

.1 After a high temperature cycle, the two inflatable lifejackets should be taken from the stowage temperature of +65°C. One should be activated using the automatic inflation system by placing it in seawater at a temperature of +30°C and the other should be activated using the manual inflation system. Each should fully inflate.

.2 After a low temperature cycle, the two inflatable lifejackets should be taken from the stowage temperature of –30°C. One should be activated using the automatic inflation system by placing it in seawater at a temperature of –1°C and the other should be activated using the manual inflation system. Each should fully inflate.

Each lifejacket should then be subjected to the tests in paragraphs 2.2, 2.3 and 2.5. A lifejacket that has been inflated automatically with one compartment uninflated should be subjected to the test in paragraph 2.2 and the test repeated until each compartment has been tested in the uninflated condition. For the fire test in paragraph 2.3, one lifejacket should be inflated and one uninflated.

2.10.1.2 After exposure to a temperature of –15°C for a period of at least 8 h, two lifejackets should be activated using the manual inflation system and should fully inflate.

2.10.1.3 After exposure to a temperature of +40°C for a period of at least 8 h, two lifejackets should be activated using the manual inflation system and should fully inflate.

2.10.2 The test in 2.7 should be conducted using lifejackets both in the inflated and uninflated conditions.

2.10.3 The tests in 2.8 should be conducted using lifejackets that have been inflated both automatically and manually, and also with one of the compartments uninflated. The tests with one of the compartments uninflated should be repeated as many times as necessary to perform the test once with each compartment in the uninflated condition.

2.10.4 *Tests of materials for inflatable bladders, inflation systems and components*

The material used for the inflatable bladder, inflation system and components should be tested to establish that they are rot-proof, colour fast and resistant to deterioration from exposure to sunlight and that they are not duly affected by seawater, oil or fungal attack.

2.10.4.1 *Material tests*

Resistance to rot and illumination tested according to AATCC Method 30:1981 and ISO 105–B04:1988 Illumination should take place to class 4-5.

Following exposure to rot or illumination tests above the tensile strength should be measured using the grab method given in ISO 5082. Minimum tensile strength should be not less than 300 N per 25 mm in the warp and weft direction.

2.10.4.2 *Coated fabrics*

Coated fabrics used in the construction of inflatable buoyancy chambers should comply with the following requirements:

.1 Coating adhesion should be tested in accordance with ISO 2411:1991 using the method described in paragraph 5.1 at 100 mm/min and should be not less than 50 N per 50 mm width.

.2 Coating adhesion should be tested when wet following ageing according to ISO 188 with an exposure of 336 ± 0.5 h in fresh water at (70 ± 1)°C, following which the method at ISO 2411:1991, paragraph 5.1 should be applied at 100 mm/min and should not be less than 40 N per 50 mm width.

.3 Tear strength should be tested in accordance with ISO 4674:1977 using method A1 and should not be less than 35 N.

.4 Resistance to flex cracking should be tested in accordance with ISO 7854:1984 method A using 9000 flex cycles; there should be no visible cracking or deterioration.

.5 Breaking strength should be tested in accordance with ISO 1421:1977 using the constant-rate-of-extension (CRE) or constant-rate-of-traverse (CRT) method, following conditioning for 24 ± 0.5 h at room temperature and should not be less than 200 N per 50 mm width.

.6 Breaking strength should be tested in accordance with ISO 1421:1977 using the CRE or CRT method, following conditioning immersed in fresh water for 24 ± 0.5 h at room temperature and should not be less than 200 N per 50 mm width.

.7 Elongation to break should be tested in accordance with ISO 1421:1977 using the CRE or CRT method following conditioning at room temperature for 24 ± 0.5 h and should not be more than 60%.

.8 Elongation to break should be tested in accordance with ISO 1421:1977 using the CRE or CRT method following conditioning immersed in fresh water at room temperature for 24 ± 0.5 h and should not be more than 60%.

.9 The resistance to exposure to light when tested in accordance with ISO 105–BO2:1988 and the contrast between the unexposed and exposed samples should not be less than class 5.

.10 The resistance to wet and dry rubbing when tested in accordance with ISO 105–X12:1995 and should not be less than class 3.

.11 The resistance to seawater should not be less than class 4 in accordance with ISO 105–EO2:1978 and the change in colour of the specimen should not be less than class 4.

2.10.4.3 *Operating head load test*

The operating head load test should be carried out using two lifejackets one lifejacket to be conditioned at −30°C for 8 h and the other at +65°C for 8 h. After mounting on the manikin or the test form the lifejackets should be inflated, and a steady force of 220 ± 10 N applied to the operating head as near as possible to the point where it enters the buoyancy chamber. This load should be maintained for 5 min during which the direction and angle in which it is applied should be continuously varied. On completion of the test the lifejacket should remain intact and should hold its pressure for 30 min.

2.10.4.4 *Pressure test*

2.10.4.4.1 *Overpressure test:* The inflatable buoyancy chambers should be capable of withstanding an internal over pressure at ambient temperature. A lifejacket should be inflated using the manual method of inflation, after inflation the relief valves should be disabled and a fully charged gas cylinder according to the manufacturer's recommendation should be fitted to the same inflation device and fired. The lifejacket should remain intact and should hold its pressure for 30 min. The lifejackets should show no signs of damage such as cracking, swelling or changes of mechanical qualities and that there has been no significant damage to the lifejacket inflation component. All fully charged gas cylinders used in this test should be sized according to the markings on the lifejacket.

2.10.4.4.2 *Relief valve test:* With one buoyancy chamber inflated, the operating head on the opposite buoyancy chamber should be fired manually, using a fully charged gas cylinder according to the manufacturer's recommendations. The operation of the relief valves should be noted to ensure that the excess pressure is relieved. The lifejacket should remain intact and should hold its pressure for 30 min. The lifejackets should show no signs of damage such as cracking, swelling or changes of mechanical qualities and that there has been no significant damage to the lifejacket inflation component.

2.10.4.4.3 *Air retention test:* One inflation chamber of a lifejacket is filled with air until air escapes from the over-pressure valve or, if the lifejacket does not have an over-pressure valve, until its design pressure, as stated in the plans and specifications, is reached. After 12 h the drop in pressure should not be greater than 10%. This test is then repeated as many times as necessary to test a different chamber until each chamber has been tested in this manner.

2.10.4.5 *Compression test*

The inflatable lifejacket, packed in the normal manner, should be laid on a table. A bag containing 75 kg of sand and having a base of 320 mm diameter should be lowered onto the lifejacket from a height of 150 mm in a time of 1 second. This should be repeated ten times, after which the bag should remain on the jacket for not less than 3 h. The lifejacket should be inflated by immersion into water and should inflate fully; the jacket to be inspected to ensure that no swelling or change of mechanical properties has occurred, the jacket should be checked for leaks.

2.10.4.6 *Test of metallic components*

2.10.4.6.1 Metal parts and components of a lifejacket should be corrosion-resistant to seawater and should be tested in accordance with ISO 9227:1990 for a period of 96 h. The metal components should be inspected and should not be significantly affected by corrosion, or affect any other part of the lifejacket and should not impair the performance of the lifejacket.

2.10.4.6.2 Metal components should not affect a magnetic compass of a type used in small boats by more than 5°, when placed at a distance of 500 mm from it.

2.10.4.7 *Inadvertent inflation test*

2.10.4.7.1 The resistance of an automatic inflation device to inadvertent operation should be assessed by exposing the entire lifejacket to sprays of water for a fixed period. The lifejacket should be fitted correctly to a free-standing manikin of adult size, with a minimum shoulder height of 1,500 mm (see figure 5), or alternatively to an appropriately sized form as shown in figure 2. The lifejacket should be deployed in the mode in which it is worn ready for use but not deployed as used in the water (i.e., if it is equipped with a cover which is normally worn closed, then the cover should be closed for the test). Two sprays should be installed so as to spray fresh water onto the lifejacket, as shown in the diagram. One should be positioned 500 mm above the highest point of the lifejacket, and at an angle of 15° from the vertical centre line of the manikin and the bottom line of the lifejacket. The other nozzle should be installed horizontally at a distance of 500 mm from the bottom line of the lifejacket, and points directly at the lifejacket. These nozzles should have a spray cone of 30°, each orifice being 1.5 ± 0.1 mm in diameter, and the total area of the orifice should be 50 ± 5 mm², the orifice being evenly spread over the spray nozzle area.

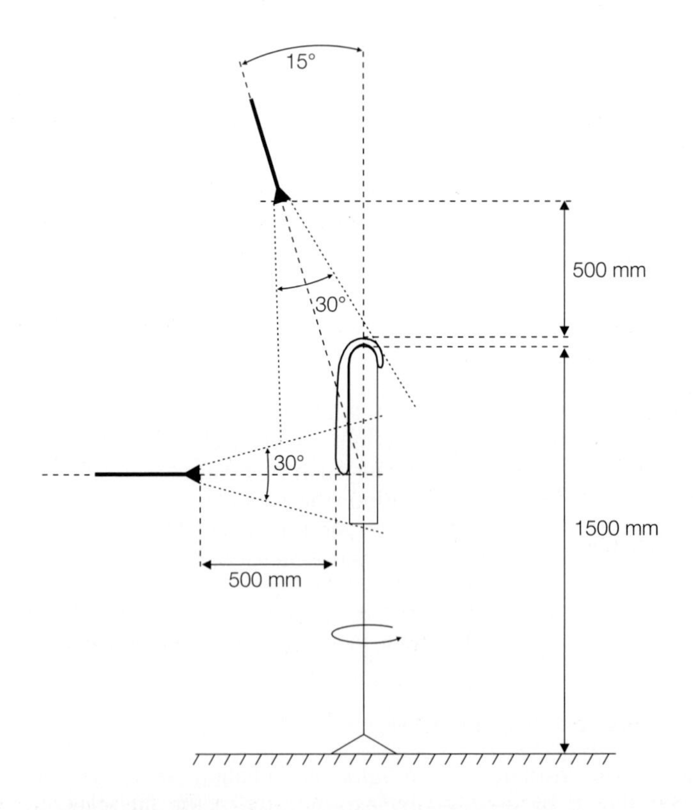

Figure 5 – *Test set-up for test of automatic inflation system*

2.10.4.7.2 The air temperature should be 20°C, and water should be supplied to the spray nozzles at a flow of 600 ℓ/h, and a temperature of 18°C to 20°C.

2.10.4.7.3 The sprays should be turned on, and the lifejacket exposed to the following series of tests to assess the ability of the jacket to resist inadvertent inflation:

.1 5 min with the high spray on the front of the lifejacket;

.2 5 min with the high spray on the left side of the lifejacket;

.3 5 min with the high spray on the back of the lifejacket; and

.4 5 min with the high spray on the right side of the lifejacket.

During exposures .1, .2 and .4, the horizontal spray should be applied for 10 periods of 3 s each to the front, left or right sides (but not back) as with the high spray.

2.10.4.7.4 After completing the above test, the lifejacket should be removed from the manikin and immersed in water to verify that the auto inflation system functions.

3 Immersion suits, anti-exposure suits and thermal protective aids

3.1 Tests common to non-insulated and insulated immersion suits and anti-exposure suits

Test subjects

3.1.1 These tests should be carried out with at least six able-bodied persons of the following heights and weights:

Height	Weight
1.4 m–1.6 m	1 person under 60 kg 1 person over 60 kg
1.6 m–1.8 m	1 person under 70 kg 1 person over 70 kg
over 1.8 m	1 person under 80 kg 1 person over 80 kg

At least one and not more than two of the persons should be females with not more than one female in the same height range.

Tests with a lifejacket

3.1.2 If the immersion suit or anti-exposure suit is to be worn in conjunction with a lifejacket, the lifejacket should be worn over the immersion suit or anti-exposure suit for the tests prescribed in 3.1.3 to 3.1.12 inclusive.

Donning test

3.1.3 Following a demonstration, each test subject should be able to unpack, don and secure the immersion suit or anti-exposure suit over their test clothing without assistance in less than 2 min. This time should include the time to don any associated clothing, inflate any orally inflated chambers if fitted, and don a lifejacket, if such is to be worn in conjunction with the immersion suit or anti-exposure suit, and the test subjects should be able to don such lifejacket without assistance.

3.1.4 The immersion suit or anti-exposure suit should be capable of being donned in 5 min at an ambient temperature as low as –30°C. Before the donning test the packed immersion suit or anti-exposure suit should be kept in a refrigerated chamber at a temperature of –30°C for 24 h.

Ergonomic test

3.1.5 When wearing the immersion suit or anti-exposure suit, the test subjects shall be able to climb up and down a vertical ladder of at least 5 m in length and demonstrate no restriction in walking, bending over or arm movement. The test subjects shall be able to pick up a pencil and write. The diameter of the pencil shall be 8 to 10 mm.

Field of vision test

3.1.6 With the heads of the seated test subjects in a fixed position, the lateral fields of vision should be at least 120° when wearing the immersion suit or anti-exposure suit.

Flotation test

3.1.7 When wearing the immersion suit or anti-exposure suit in conjunction with a lifejacket if required, the test subjects should float face-up with their mouths clear of the water by at least 120 mm and be stable in that position. For a buoyant insulated immersion suit worn without a lifejacket, an auxiliary means of buoyancy such as an orally inflated bladder behind the wearer's head may be used to obtain this freeboard, provided that the freeboard obtained without the auxiliary means of buoyancy is at least 50 mm. The freeboard should be measured from the water surface to the nose and mouth with the test subject at rest. The freeboard of the anti-exposure suit without a lifejacket should be at least 50 mm. The position of the lifejacket light should permit it to be visible over as great a segment of the upper hemisphere as is practicable.

Righting test

3.1.8 Except where it has been demonstrated that the immersion suit or anti-exposure suit will right the test subjects within 5 s, the test subjects should each demonstrate that they can turn themselves from a face-down to a face-up position in not more than 5 s.

Water ingress and jump test

3.1.9 Following a jump by each test subject into water from a height sufficient to totally immerse the body, the ingress of water into the immersion suit or anti-exposure suit should not exceed a mass of 500 g. This may be determined from the difference in the combined mass of the test subject and the immersion suit or anti-exposure suit (pre-wetted), as measured prior to the jump and immediately after the jump. Weighings should be performed on a machine accurate to ± 100 g.

3.1.10 The immersion suit or anti-exposure suit and its attachments should not be damaged or dislodged in any way following a jump from a height of 4.5 m vertically into the water. It should be established by questioning the test subjects that the suit does not injure the wearer as a result of this test.

Leak test

3.1.11 The ingress of water into the pre-wetted suit should not exceed a mass of 200 g following:

.1 a period of flotation in calm water for 1 h; or

.2 swimming for 20 min for a distance of at least 200 m.

The mass of water ingress should be measured by weighing the test subject and the suit in accordance with the method prescribed in 3.1.9.

Swimming and water emergence test

3.1.12 All test subjects, each wearing a lifejacket but not the immersion suit or anti-exposure suit, should attempt to swim 25 m and board a liferaft or a rigid platform with its surface 300 mm above the water surface. Test subjects who successfully complete this task should also perform it wearing the immersion suit or anti-exposure suit.

Tests for oil resistance

3.1.13 After all its apertures have been sealed, an immersion suit or anti-exposure suit should be immersed under a 100 mm head of diesel oil for 24 h. The surface oil should then be wiped off and the immersion suit or anti-exposure suit subjected to the test prescribed in 3.1.11. The ingress of water should not exceed a mass of 200 g.

3.1.14 In lieu of the test for oil resistance prescribed in 3.1.13, either of the following tests may be conducted:

.1 After all apertures have been sealed, the suit should be immersed under a 100 mm head of diesel oil for a period of 24 h at normal room temperature, if necessary using weights to keep the suit submerged. Any surface oil should then be wiped off and the suit turned inside out. The suit should then be laid on a table suitable for collecting and draining off any leakage and be supported at the neck aperture by a suitably designed hanger. The suit should then be filled with water to neck level which should be 300 mm above the table. The suit should be left in this position for 1 h and the leakage collected and weighed. The leakage should not exceed a mass of 200 g.

.2 Representative samples of the exterior fabric and seams should be immersed under a 100 mm head of diesel oil for 24 h. After removal from the oil, samples should be wiped off before being subjected to a hydrostatic test of a 1 m water head and a tensile seam strength of 150 N.

Fire test

3.1.15 An immersion suit or anti-exposure suit should be subjected to the fire test as prescribed in 1.5. If necessary, the immersion suit or anti-exposure suit should be draped over a hanger to ensure the whole immersion suit or anti-exposure suit is enveloped in the flames. The immersion suit or anti-exposure suit should not sustain burning for more than 6 s or continue melting after being removed from the flames.

Temperature cycling test

3.1.16 An immersion suit or anti-exposure suit should be subjected to the temperature cycling as prescribed in 1.2.1 and should show no sign of damage such as shrinking, cracking, swelling, dissolution or change of mechanical qualities.

Buoyancy test

3.1.17 A buoyancy test, as prescribed in 2.2, should be carried out to establish that the buoyancy of an immersion suit or anti-exposure suit designed to be worn without a lifejacket is not reduced by more than 5% after 24 h submersion in fresh water.

Strength test

3.1.18 The immersion suit or anti-exposure suit should be subjected to the body strength tests prescribed in 2.6.1, except the load applied should be 1,350 N. The immersion suit or anti-exposure suit may be cut if necessary to accommodate the test device.

3.2 Thermal protective tests

General

3.2.1 These tests should be performed as described below. The thermal protective qualities may be measured using a thermal manikin, when such a method is required by an Administration and has been demonstrated to provide test results which correlate satisfactorily in all aspects to test results using human subjects.

3.2.2 If the test should be performed by human subjects, they should be medically examined before being accepted for participation in the tests. Each design of immersion suit- or anti-exposure suit is to be tested by the test subjects specified in 3.1.1

3.2.3 Where human subjects are used, the tests should always be conducted under the supervision of a physician. Emergency resuscitation equipment should be available during all tests. For safety reasons, ECG should be monitored during every test. Testing should be stopped at the wish of the test subjects, if the falling rate of the core temperature is more than 1.5°C per hour after the first half hour, if the skin temperature of the hand, foot or lumbar region should fall below 10°C, or if the attending physician considers it advisable.

3.2.4 When testing with human subjects, continuous body core temperature (rectal temperature) and skin temperatures of lumbar region, both hands, calves, feet (foot instep) and heels, should be measured. The accuracy of the measuring system should be ± 0.2°C. Appropriate corresponding measurements should be taken if a manikin is used in lieu of human subjects.

3.2.5 Prior to the tests, the same amount of water resulting from the jump test in 3.1.9 should be poured into the dry immersion suit or anti-exposure suit worn over the dry test clothing specified in 3.2.6 by the test subject lying down.

Test clothing

3.2.6 The test subjects should wear a standard range of clothing consisting of:

.1 underwear (short sleeved, short legged);

.2 shirt (long sleeved);

.3 trousers (not woollen); and

.4 woollen socks.

3.2.7 If the immersion suit or anti-exposure suit is to be worn in conjunction with a lifejacket, the lifejacket should be worn during the thermal protective tests.

Specific tests for non-insulated immersion suits

3.2.8 In addition to the clothing specified in 3.2.6 and 3.2.7, the test subject should wear two woollen pullovers.

3.2.9 Each test subject should wear an immersion suit previously subjected to the jump test in 3.1.10. Following a 1 h period of immersion, with the hands gloved, in circulating calm water at +5°C, each test subject's body core temperature should not fall more than 2°C below the normal level of the test subject's temperature.

3.2.10 Immediately on leaving the water after completion of the test prescribed in 3.2.9 the test subject should be able to pick up a pencil as specified in 3.1.5 and write.

Specific tests for insulated immersion suits

3.2.11 Each test subject should wear an immersion suit previously subjected to the jump test in 3.1.10. Following a 6 h period of immersion, with the hands gloved, in circulating calm water at between 0°C and +2°C, each test subject's body core temperature should not fall more than 2°C below the normal level of the test subject's temperature.

3.2.12 The immersion suit should provide sufficient thermal protection to ensure that immediately on leaving the water after a 1 h period of immersion, with hands gloved, in circulating calm water at +5°C, each test subject can pick up a pencil as specified in 3.1.5 and write. Alternatively, at the manufacturer's option, the ability to pick up a pencil as specified in 3.1.5 and write may be demonstrated immediately on leaving the water after completion of the test prescribed in 3.2.11.

Specific tests for anti-exposure suits

3.2.13 Each test subject should wear an anti-exposure suit previously subjected to the jump test in 3.1.10. Following a 1 h period of immersion, with the hands gloved and hood donned, in circulating calm water at a temperature of +5°C, each test subject's body core temperature should not fall more than 2°C below the normal level of the test subject's temperature.

3.2.14 Immediately on leaving the water after completion of the test prescribed in 3.2.13, the test subject should be able to pick up a pencil as specified in 3.1.5 and write.

3.3 Thermal protective aids for survival craft

Fabric test

3.3.1 It should be demonstrated that the fabric from which the thermal protective aid is constructed can maintain its watertight integrity when supporting a column of water 2 m high.

3.3.2 It should be demonstrated by test that the fabric has a thermal conductivity of not more than 7,800 $W/(m^2 \cdot K)$.

Temperature cycling test

3.3.3 A thermal protective aid should be subjected to temperature cycling as prescribed in 1.2.1 and should show no sign of damage such as shrinking, cracking, swelling, dissolution or change of mechanical qualities.

Test subjects

3.3.4 For these tests a group of at least six test subjects of different ages, both male and female in the large, medium and small size range should be selected.

Test clothing

3.3.5 The test clothing worn by the test subjects should be as prescribed in 3.2.6 and 3.2.8.

Donning test

3.3.6 Following a demonstration, the test subjects should be able to unpack and don the thermal protective aid over a lifejacket when seated in a survival craft.

3.3.7 The thermal protective aid should be capable of being unpacked and donned at an ambient temperature of –30°C. Before the donning test the thermal protective aid should be kept in a refrigerated chamber at a temperature of –30°C for 24 h.

Discarding test

3.3.8 If the thermal protective aid impairs the ability of the test subjects to swim, it should be demonstrated that it can be discarded by the test subjects, when immersed in water, in not more than 2 min.

Test for oil resistance

3.3.9 After all its apertures have been sealed, a thermal protective aid should be immersed under a 100 mm head of diesel oil for 24 h. The surface oil should then be wiped off and it should be established that the thermal conductance is not more than 7,800 $W/(m^2 \cdot K)$.

4 Pyrotechnics – rocket parachute flares, hand flares and buoyant smoke signals

4.1 General

A minimum of three specimens of each type of pyrotechnic should be subjected to each individual test. All three specimens should pass each individual test.

4.2 Temperature tests

Three specimens of each type of pyrotechnic should be subjected to:

.1 temperature cycling as prescribed in 1.2.1. After the test each specimen should show no sign of damage such as shrinking, cracking, swelling, dissolution or change of mechanical qualities and then function effectively at ambient temperature;*

* Not applicable to smoke signals, for which see paragraphs 1.9.2 and 4.8.1.

.2 a temperature of –30°C for at least 48 h and then function effectively immediately upon removal from the cold chamber;*

.3 a temperature of +65°C for at least 48 h and then function effectively immediately upon removal from the hot chamber;*

.4 a temperature of +65°C and 90% relative humidity for at least 96 h, followed by ten days at 20°C to 25°C at 65% relative humidity and then function effectively.

4.3 Water and corrosion resistance test

Nine specimens of each type of pyrotechnic should function effectively after being subjected to the following tests (three specimens to each test):

.1 immersed horizontally for 24 h under 1 m of water;

.2 immersed in the ready-to-fire condition for 5 min under 10 cm of water;

.3 subjected to a salt spray (5% natrium chloride solution) at a temperature of +35 ± 3°C for at least 100 h.

4.4 Handling safety test

Three specimens of each type of pyrotechnic should:

.1 be dropped in turn end-on and horizontally from a height of 2 m onto a steel plate about 6 mm thick cemented on to a concrete floor. It should remain in a safe condition after this test and should subsequently be operated and function effectively; and

.2 be activated in accordance with the manufacturer's operating instructions by an operator wearing an insulated buoyant immersion suit or the gloves taken from an insulated buoyant immersion suit, to establish that it can be operated effectively without injury to the operator, or any person in close proximity, during firing or burning.

4.5 Safety inspection

It should be established by visual inspection that each type of pyrotechnic:

.1 is indelibly marked with clear and precise instructions on how it should be operated and that the danger end can be identified by day or night;

* Not applicable to smoke signals, for which see paragraphs 1.9.2 and 4.8.1

.2 can, if hand-operated, be operated from the bottom (safe end) or that it contains an operational safety delay of 2 s;

.3 has, in the case of a rocket parachute flare and hand flare, an integral means of ignition;

.4 has a simple means of ignition which requires the minimum of preparation and can be readily operated in adverse conditions without external aid and with wet, cold or gloved hands;

.5 does not depend on adhesive tapes or plastic envelopes for its water-resistant properties; and

.6 can be indelibly marked with means for determining its age.

4.6 Rocket parachute flares test

4.6.1 Three rockets should be fired vertically. After firing it should be established by means of accurate measuring instruments that the parachute flare is ejected at a height of not less than 300 m. The height at which the flare burns out and the burning period should also be measured. It should be established from these measurements that the rate of descent is not more than 5 m/s and the burning period is not less than 40 s.

4.6.2 Laboratory testing of the flare material should establish that it will burn uniformly with an average luminous intensity of not less than 30,000 cd and that the colour of the flame is a vivid red with CIE co-ordinates $x = 0.61$ to 0.69 and $y = 0.3$ to 0.39, or computed from these co-ordinates: a wavelength of 608 ± 11 nm.

4.6.3 Three rockets should function efficiently when tested by firing at an angle of 45° to the horizontal.

4.6.4 If the rocket is hand-held when operated, it should be demonstrated that its recoil is minimal.

4.7 Hand flares test

4.7.1 Three flares should be activated and should burn for a period of not less than 1 min. After burning for 30 s, each flare should be immersed under 100 mm of water for a period of 10 s and should continue burning for at least a further 20 s.

4.7.2 Laboratory testing of the flare material should establish that it will burn with an average luminous intensity of at least 15,000 cd and that the colour of the flame is vivid red with CIE co-ordinates $x = 0.61$ to 0.69 and $y = 0.3$ to 0.39, or computed from these co-ordinates: a wavelength of 608 ± 11 nm.

4.7.3 Three flares should be activated 1.2 m above a test pan 1 m square containing 2 ℓ of heptane floating on a layer of water. The test should be conducted at an ambient temperature of +20°C to +25°C. The flare should be allowed to burn completely and the heptane should not be ignited by the flare or material from the flare.

4.8 Buoyant smoke signals test

4.8.1 Nine buoyant smoke signals should be subjected to temperature cycling as prescribed in 1.2.1. After at least ten complete temperature cycles, three smoke signals should be taken from a stowage temperature of –30°C, be activated and should then operate in seawater at a temperature of –1°C. The next three smoke signals should be taken from a stowage temperature of +65°C, be activated and should then operate in seawater at a temperature of +30°C. The last three smoke signals should be taken from ordinary room conditions and activated. After emitting smoke for 1 min, they should be fully submerged for a period of not less than 10 s and should continue emitting smoke during and after submersion and demonstrate a total period of smoke emission of not less than 3 min.

4.8.2 Three smoke signals should function in water covered by a 2 mm layer of heptane without igniting the heptane.

4.8.3 The smoke density and colour of the smoke signal should be determined by laboratory testing conducted at a water temperature of +20°C to +25°C as follows:

.1 The smoke should be blown through an apparatus consisting of a 190 mm diameter duct with a fan capable of producing an entrance air flow of 18.4 m^3/min. By means of a light source with at least 10 cd on one side of the tunnel and a photoelectric cell on the other side, the density of the passing smoke should be recorded. If the photocell picks up the total emitted light from the light source, then the smoke density is zero percent which means that no smoke is passing through the tunnel. The smoke density is then considered to be 100% when the photocell is not able to pick up any light of the light source through the passing smoke in the tunnel. From the amount of light which the photocell is able to pick up the smoke density should be calculated. Before each measurement, the light intensity of the 100% value should be checked. Each measurement should be recorded.

.2 The colour of the orange smoke should be evaluated by means of visual comparison, in daylight, to a colour comparison chart containing the range of acceptable orange colours. The colour comparison chart should have a gloss or matte finish, and consist of a series of at least five orange colour chips, covering the range from reddish orange (Munsell notation 8.75 YR 6/14) to yellowish orange (Munsell notation 5 YR MAX) in gradual steps of hue, chroma, and lightness. The colour chips should be secured adjacent to one another, in order of progression from reddish orange to yellowish orange, and extend on at least one side to the edge of the chart. Each colour chip should be at least 50 mm × 100 mm in size.

Notes: .1 A typical acceptable progression would be 8.75 YR 6/14; 10 R 6/14; 1.25 YR 6/14; 3.75 YR MAX; 5 YR MAX.

.2 ASTM D1535-97 specifies a method to convert between Munsell notation and CIE co-ordinates.

4.8.4 A smoke signal should be tested in waves at least 300 mm high. The signal should function effectively for not less than 3 min.

5 Liferafts – rigid and inflatable

5.1 Drop test

5.1.1 Each type of liferaft should be subjected to a minimum of two drop tests. Where the liferaft in its operational condition is packed in a container or valise, one such test should be carried out with the liferaft packed in each type of container or valise in which the manufacturer proposes to market it.

5.1.2 The liferaft, in the operationally packed condition, should be suspended and then dropped from a height of 18 m into the water. If it is to be stowed at a height greater than 18 m, it should be dropped from the height at which it is to be stowed. The free end of the painter should be attached to the point of suspension so that it pays out as the liferaft drops, thus simulating actual conditions.

5.1.3 The liferaft should be left floating for 30 min:

.1 in the case of a rigid liferaft it should be lifted from the water to permit thorough inspection of the liferaft, the contents of the equipment container and, where applicable, the container or valise; and

.2 in the case of an inflatable liferaft, it should then be inflated. The liferaft should inflate upright and in the time prescribed in 5.17.3 to 5.17.6. The thorough inspection prescribed in 5.1.3.1 should then be carried out.

5.1.4 Damage to the container or valise, if the liferaft is normally within it when launched, is acceptable provided the Administration is satisfied that it would not be a hazard to the liferaft. Damage to any item of equipment is acceptable subject to the Administration being satisfied that the operational efficiency has not been impaired. Damage to fresh water receptacles may be accepted provided they do not leak. However, for drop tests from heights exceeding 18 m, leakage from up to 5% of the receptacles may be accepted provided that:

.1 the equipment list for the inflatable liferaft specifies the carriage of 5% excess water or means of desalination adequate to produce an equivalent amount; or

.2 the water receptacles are contained in a waterproof overwrap.

5.2 Jump test

5.2.1 It should be demonstrated that a person can jump onto the liferaft, with and without the canopy erected, from a height above the floor of at least 4.5 m without damaging the liferaft. The test subject should weigh not less than 75 kg and should be wearing hard bottom shoes with smooth soles and no protruding nails. The number of jumps performed should be equal to the total number of persons for which the liferaft is to be approved.

5.2.2 The jump test may be simulated by dropping a suitable and equivalent mass, arranged so as to impact the liferaft with shoes as described in 5.2.1.

5.2.3 There should be no torn fabric or damage to seams as a result of the test.

5.2.4 Unless the configuration of both sides of a canopied reversible liferaft are identical, this test should be repeated for both sides of the liferaft.

5.3 Weight test

The fully packed liferaft container should be weighed to determine whether its mass exceeds 185 kg. The weight test should be performed on the heaviest variation of the liferaft, considering different containers and equipment packs which may be used. If the mass exceeds 185 kg, the different combinations of containers and equipment packs should be weighed to determine which will and which will not exceed 185 kg.

5.4 Towing test

It should be demonstrated by towing that the fully loaded and equipped liferaft is capable of being satisfactorily towed at speeds of up to 3 knots in calm water. Towing should be by a line attached to the liferafts towing connection. The sea anchor should be streamed while the liferaft is towed. The liferaft should be towed for a distance of at least 1 km. During the test the force required to tow the liferaft should be measured at speeds of 2 knots and 3 knots and recorded on the type approval certificate.

5.5 Mooring out tests

The liferaft should be loaded with mass equal to the mass of the total number of persons for which it is to be approved and its equipment and moored in a location at sea or in a seawater harbour. The liferaft should remain afloat in that location for 30 days. In the case of an inflatable liferaft, the pressure may be topped up once a day using the manual pump; however, during any 24 h period the liferaft should retain its shape. The liferaft should not sustain any damage that would impair its performance. After this test, the inflatable liferaft should be subjected to the pressure test prescribed in 5.17.7 and 5.17.8.

5.6 Liferaft painter system test

The breaking strength of the painter system including its means of attachment to the liferaft should be as follows:

.1 not less than 7.5 kN for liferafts accommodating up to 8 persons;

.2 not less than 10.0 kN for liferafts accommodating 9 to 25 persons;

.3 not less than 15.0 kN for liferafts accommodating more than 25 persons.

5.7 Loading and seating test

The freeboard of the liferaft in the light condition, including its full equipment but no personnel, should be recorded. The freeboard of the liferaft should again be recorded when the number of persons for which the liferaft is to be approved, having an average mass of 75 kg, and each wearing an immersion suit and lifejacket, have boarded and are seated. It should be established that all the seated persons have sufficient space and headroom and it should be demonstrated that the various items of equipment can be used within the liferaft in this condition and, in the case of an inflated liferaft, with the floor inflated. The freeboard, when loaded with the mass of the number of persons for which it is to be approved and its equipment, with the liferaft on an even keel and, in the case of an inflatable liferaft, with the floor not inflated, should not be less than 300 mm. Unless the configuration of both sides of a canopied reversible liferaft are identical, this test should be repeated for both sides of the liferaft.

5.8 Boarding and closing arrangement test

The boarding test should be carried out in a swimming pool by a team of not more than four persons who should be of mature age and of differing physiques as determined by the Administration. Preferably they should not be strong swimmers. For this test they should be clothed in shirt and trousers or a boiler suit and should wear approved lifejackets suitable for an adult. They must each swim about 100 m before reaching the liferaft for boarding. There must be no rest period between the swim and the boarding attempt. Boarding should be attempted by each person individually with no assistance from other swimmers or persons already in the liferaft. The water should be of a depth sufficient to prevent any external assistance when boarding the liferaft. The arrangements will be considered satisfactory if three of the persons board the liferaft unaided and the fourth boards with the assistance of any of the others. The above-mentioned test should be carried out also with persons clothed in immersion suits and lifejackets. After the boarding test, it should be demonstrated by a person clothed in an approved immersion suit that the canopy entrance can be easily and quickly closed in 1 min and can be easily and quickly opened from inside and outside in 1 min. Unless the configuration of both sides of a canopied reversible liferaft are identical, this test should be repeated for both sides of the liferaft.

5.9 Stability test

5.9.1 The number of persons for which the liferaft is to be approved should be accommodated on one side and then at one end and in each case the freeboard should be recorded. Under these conditions the freeboard should be such that there is no danger of the liferaft being swamped. Each freeboard measurement should be taken from the waterline to the top surface of the uppermost main buoyancy tube at its lowest point.

5.9.2 The stability of the liferaft during boarding may be ascertained as follows: two persons, each wearing approved lifejackets, should board the empty liferaft. It should then be demonstrated that the two persons in the liferaft can readily assist from the water a third person who is required to feign unconsciousness. The third person must have his back towards the entrance so that he cannot assist the rescuers. It should be demonstrated that the water pockets adequately counteract the upsetting moment on the liferaft and there is no danger of the liferaft capsizing.

5.10 Manoeuvrability test

It should be demonstrated that with the paddles provided, the liferaft is capable of being propelled when fully laden in calm conditions over a distance of at least 25 m.

5.11 Swamp test

It should be demonstrated that the liferaft is fully swamped, it is capable of supporting the number of persons for which it is to be approved and remains seaworthy. The liferaft should not seriously deform in this condition. The swamped inflatable liferaft should be tested in at least 10 waves at least 0.9 m high. The waves may be produced by the wake of a boat, or by other acceptable means. During this test self-draining arrangements fitted in the floor of the liferaft are to be closed to prevent the ingress of water. Unless the configuration of both sides of a canopied reversible liferaft are identical, this test should be repeated for both sides of the liferaft.

5.12 Canopy closure test

To ensure the effectiveness of the canopy closures in preventing water entering the liferaft, the efficiency of the closed entrances should be demonstrated by means of a hose test or by any other equally effective method. The requirement for the hose test is that about 2,300 ℓ of water per minute be directed at and around the entrances through a 63.5 mm hose from a point 3.5 m away and 1.5 m above the level of the buoyancy tubes for a period of 5 min. The accumulation of water inside the liferaft should not exceed 4 ℓ. Unless the configuration of both sides of a canopied reversible liferaft are identical, this test should be repeated for both sides of the liferaft.

5.13 Buoyancy of float-free liferafts

It should be demonstrated that the liferafts packed in containers which are float-free have sufficient inherent buoyancy to inflate the liferaft by means of the actuating line in the event of the ship sinking. The combination of equipment and container or valise should be that which produces the maximum packed weight.

5.14 Detailed inspection

A liferaft, complete in all respects and, if an inflatable liferaft, in a fully inflated condition should be subjected to a detailed inspection in the manufacturer's works to ensure that all the Administration's requirements are fulfilled.

5.15 Weak link test

The weak link in the painter system should be tensile tested and should have a breaking strain of 2.2 ± 0.4 kN.

5.16 Davit-launched liferafts – strength test of lifting components

5.16.1 The breaking strength of the webbing or rope and the attachments to the liferaft used for the lifting bridle should be established by tests on three separate pieces of each different item. The combined strength of the lifting bridle components should be at least six times the mass of the liferaft when loaded with the number of persons for which it is to be approved and its equipment.

Impact test

5.16.2 The liferaft should he loaded with a mass equal to the mass of the number of persons for which it is to be approved and its equipment. With the liferaft in a free hanging position it should be pulled laterally to a position so that when released it will strike a rigid vertical surface at a velocity of 3.5 m/s. The liferaft should then be released to impact against the rigid vertical surface. After this test the liferaft should show no signs of damage which would affect its efficient functioning.

Drop test

5.16.3 The liferaft, loaded as prescribed in 5.16.2, should be suspended from an on-load release at a height of 3 m above the water, be released and allowed to fall freely into the water. The liferaft should then be examined to ensure that no damage has been sustained which would affect its efficient functioning.

Davit-launched liferaft boarding test

5.16.4 A davit-launched liferaft should, in addition to the boarding test prescribed in 5.8, be subjected to the following test. The liferaft, should be suspended from a liferaft launching appliance, or from a crane with a head sheave of similar height, and bowsed into the ship's side or simulated ship's side. The liferaft should then be boarded by the number of persons for which it is to be approved of average mass 75 kg. There should be no undue distortion of the liferaft. The bowsing should then be released and the liferaft left hanging for 5 min. It should then be lowered to the sea or floor and unloaded. At least three tests are required in succession, with the hook of the lowering appliance so positioned that its distance from the ship's side is:

.1 half the beam of the liferaft +150 mm;

.2 half the beam of the liferaft; and

.3 half the beam of the liferaft –150 mm.

The boarding, which is intended to simulate actual shipboard conditions, should be timed and the time recorded.

5.17 Additional tests applicable to inflatable liferafts only

Damage test

5.17.1 It should be demonstrated that, in the event of any one of the buoyancy compartments being damaged or failing to inflate, the intact compartment or compartments should support, with positive freeboard over the liferaft's periphery, the number of persons for which the liferaft is to be approved. This can be demonstrated with persons each having a mass of 75 kg and seated in their normal positions or by an equally distributed mass.

Righting test

5.17.2 This test is not required for a canopied reversible liferaft. For this test the liferaft should be inverted so as to simulate inverted inflation.

.1 The inflatable liferaft should be loaded with its heaviest equipment pack. All of the entrances, ports, and other openings in the liferaft canopy should be open in order to allow the infiltration of water into the canopy when capsized;

.2 the canopy of the liferaft should then be completely filled with water. Except for an automatically self-righting liferaft this may be accomplished, if necessary, by partially collapsing the canopy support, or alternatively the uninflated liferaft should be flaked out onto the surface of the water upside down and inflation initiated. An automatically self-righting liferaft should self-right in this condition, and should become boardable in the upright position within 1 min after the start of the test. If the inflatable liferaft, other than an automatically self-righting liferaft, does not self-right, it should be allowed to remain in an inverted position for at least 10 min before righting is attempted;

.3 the righting test should be carried out by the same team of persons required for the boarding test similarly clothed and wearing lifejackets and after completing the swim required in 5.8. At least one of the persons righting the inflatable liferaft should weigh less than 75 kg. Each person should attempt to right the liferaft unaided. The water should be of sufficient depth to give no external assistance to the swimmers when mounting the inverted liferaft;

.4 the righting arrangements will be considered satisfactory if each person rights the liferaft unaided. There should be no damage to the structure of the inflatable liferaft, and the equipment pack should remain secured in its place.

Inflation test

5.17.3 A liferaft, packed in each type of container, should be inflated by pulling the painter and the time recorded:

.1 for it to become boardable, i.e., when buoyancy tubes are inflated to full shape and diameter;

.2 for the cover to be erect; and

.3 for the liferaft to reach its full operational pressure* when tested:

.3.1 at an ambient temperature of between 18°C and 20°C;

.3.2 at a temperature of –30°C; and

.3.3 at a temperature of +65°C.

5.17.4 When inflated in an ambient temperature of between 18°C and 20°C, it should achieve total inflation in not more than 1 min. In the case of automatic self-righting liferaft, the liferaft should achieve total inflation and be boardable in the upright position in not more than 1 min, regardless of the

* The term *operational pressure* has the same meaning as the term *working pressure*; i.e., the pressure determined by the designed reseat pressure of the relief valves, if fitted, except that, if the actual reseat pressure of the relief valves, determined by testing, exceeds the designed reseat pressure by more than 15%, the higher figure should be used.

orientation in which the liferaft inflates. The force required to pull the painter and start inflation should not exceed 150 N.

5.17.5 For the inflation test at –30°C the packed liferaft should be kept at room temperature for at least 24 h, then placed in a refrigerated chamber at a temperature of –30°C for 24 h prior to inflation by pulling the painter. Under these conditions the liferaft should reach working pressure in 3 min. Two liferafts should be subject to an inflation test at this temperature. There should be no seam slippage, cracking, or other defect in the liferaft and it should be ready for use after the tests.

5.17.6 For the inflation test at +65°C the packed liferaft should be kept at room temperature for at least 24 h, then placed in a heating chamber at a temperature of +65°C for not less than 7 h prior to inflation by pulling the painter. Under these conditions the gas pressure relief valves must be of sufficient capacity to prevent damage to the liferaft by excess pressure and to prevent the maximum pressure during the inflation from reaching twice the reseat pressure of the release valve. There must be no seam slippage, cracking or other defect in the liferaft.

Pressure test

5.17.7 Each inflatable compartment in the liferaft should be tested to a pressure equal to three times the working pressure. Each pressure relief valve should be made inoperative, compressed air should be used to inflate the inflatable liferaft and the inflation source removed. The test should continue for at least 30 min. The pressure should not decrease by more than 5% as determined without compensating for temperature and atmospheric pressure changes, and there should be no seam slippage, cracking or other defect in the liferaft.

5.17.8 The measurement of pressure drop due to leakage can be started when it has been assumed that compartment material has completed stretching due to the inflation pressure and achieved equilibrium.

Seam strength test

5.17.9.1 It should be demonstrated that sample seams, prepared in the same condition as in production, can withstand a test load equal to the minimum specified liferaft fabric tensile strength. Sewn seams on outer canopy fabric should withstand a test load of at least 70% of the minimum specified fabric tensile strength when tested by the method described in ISO 1421 and by using test samples as shown in figure 1 below.

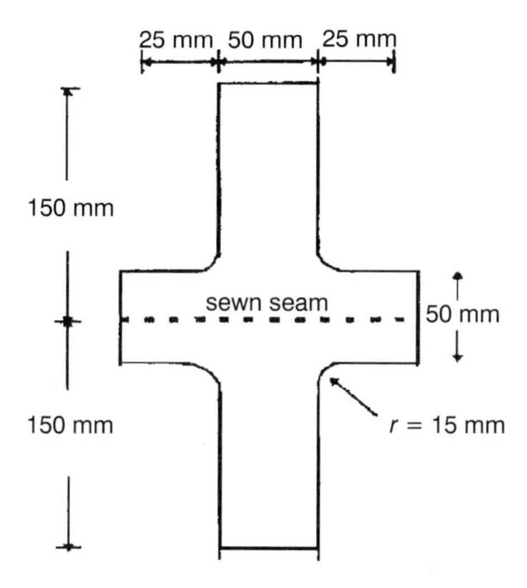

Figure 1 – *Sample specification for sewn canopy seams*

Samples of all types of sewing used in production should be tested.

Seam constructions in both warp and weft direction should be tested.

The test specimens should be cut out from pre-sewn samples of fabric and no locking of thread ends should take place.

5.17.9.2 *Weld strength*

.1 When tested by the method prescribed below, the load required to initiate failure of the weld should be not less than 175 N;

.2 Specimens should be prepared and tested as given in .3 below:

5.17.9.3 Hydrolysis tests should be conducted on sample welded seams where thermoplastic coated materials are to be used. The tests should be conducted as follows:

.1 When tested by the method prescribed below, the weld strength of the sample seam should achieve 125 N/25 mm minimum.

.2 Test method:

.2.1 Store the test specimens for 12 weeks over water in a closed container at 93 ± 2°C.

.2.2 After the conditioning as above, dry the specimens for 1 h at 80 ± 2°C and condition at 20 ± 2°C, 65% RH for 24 h.

.3 Welded test samples should be prepared as follows:

Two samples of fabric 300 mm × 200 mm, cut with the short side parallel to the warp direction, should be superimposed face to back for double coated fabrics, or coated face to coated face for single or asymmetrically coated fabrics. They should be welded with a tool 10 ±1 mm width of convenient length. 25 mm wide test specimens should be cut transversely to the line of the weld. The test samples should be mounted in a test machine as in ISO 1421. The maximum peel load should be recorded.

Davit-launched inflatable liferafts – strength test

5.17.10 It should be demonstrated by an overload test on the liferaft hanging from its centre support that the bridle system has an adequate factor of safety as follows:

.1 the liferaft should be placed in a temperature of 20 ± 3°C for a period of at least 6 h;

.2 following this period of conditioning, the liferaft should be suspended from its lifting hook or bridle and the buoyancy chambers (not including an inflatable floor) inflated;

.3 when fully inflated and when the relief valves have reseated themselves, all relief valves should be made inoperative;

.4 the liferaft should then be lowered and loaded with a distributed mass equivalent to four times the mass of the number of persons for which it is to be approved and its equipment, the mass of each person being taken as 75 kg;

.5 the liferaft should then be raised and remain suspended for at least 5 min;

.6 the pressure before and after the test after the weight is removed and while it remains suspended, should be recorded; and

.7 any dimensional deflections or distortions of the liferaft should be recorded. During the test and after its completion, the inflatable liferaft should remain suitable for its intended use.

5.17.11 It should be demonstrated, after a period of 6 h in a chamber at a temperature of –30°C, that the liferaft will support a load of 1.1 times the number of persons for which it is to be approved and its equipment with all relief valves operative. The liferaft should be loaded with the test weight in the refrigerated chamber. The floor should not be inflated. The loaded inflatable liferaft should remain suspended for at least 5 min. If the inflatable liferaft must be removed from the chamber in order to suspend it, the inflatable liferaft should be suspended immediately upon removal from the chamber. During the test and after its completion, the inflatable liferaft should remain suitable for its intended use.

5.17.12 The inflatable liferaft should be loaded with a weight equal to the mass of its heaviest equipment pack and the number of persons for which it is to be approved, the mass of each person being taken as 75 kg. Except for the floor which should not be inflated, the inflatable liferaft should be fully inflated with all relief valves operative. A liferaft should be lowered for a distance of at least 4.5 m in continuous contact against a structure erected to represent the side of a ship having a 20° adverse list. The height of the point from which the hook is suspended should be comparable to that of a shipboard launching appliance. During the test and after its completion, the liferaft should not sustain damage or distortion, or assume a position which would render it unsuitable for its intended purpose.

Material tests

5.17.13 The materials when tested should meet the following requirements:

.1 The fabric should be marked in such a manner as to allow traceability of the fabric manufacturer and production lot number.

.2 Tests and performance criteria:

.2.1 Test samples should be randomly selected; and test pieces cut from each sample as required by the relevant ISO standard or as otherwise prescribed for each individual test.

.2.2 Fabric used for manufacture of buoyancy tubes, inflatable supports for canopies and floors should meet the following requirements:

.2.2.1 *Tensile strength*

When tested by the method described in ISO 1421 the tensile strength should be a minimum of 2,255 N/50 mm width for warp and weft. Maximum elongation, for the above should be 30% over a 200 mm gauge length, the elongation should be expressed as a percentage of the initial test length between the jaws. Where two layers of floor fabric are provided to form an inflatable floor the main floor should be as specified. The inner/outer layer may have a minimum tensile strength of 1,470 N/50 mm widths in warp and weft direction.

.2.2.2 *Tear strength*

When tested with the apparatus described in ISO 1421, the tear strength should be 1,030 N minimum, in warp and weft directions. Where two layers of the floor fabric are provided to form an inflatable floor, the main floor should be as specified. The inner/outer layer may have a minimum tear strength of 735 N in warp and weft directions. The preparation of the test specimens should be as follows:

.1 from the test sample cut three specimens each in warp and weft directions, 76 ±1 mm wide and 400 mm long, with the length closely parallel to the warp and weft yarns. Space the selection across the full length and width of the sample. Make a 12.5 mm cut across the middle of each specimen at right angles to the length.

.2 grip the specimen under test securely and evenly in the grips, which should be 200 mm apart, so that the specimen length is closely in the direction of the pull. Operate the machine in accordance with ISO 1421. The maximum load sustained is recorded as the wound tear strength, and the average for the three specimens is calculated.

.2.2.3 *Surface receptiveness and adhesion of surface coating*

.1 When tested by the method described in ISO 2411 the surface receptiveness on either face should not be less than 75 N/50 mm width.

.2 For dry surface coating adhesion a minimum of 75 N/50 mm is required.

.3 For wet surface coating adhesion as described in 5.17.13.2.2.3.8 a minimum of 50 N/50 mm is required.

.4 Each coated face should be tested. The specimens should be made up as in ISO 2411 bonding like-coated face to like-coated face.

.5 The bonding used and the method of application should be agreed between the liferaft manufacturer and the finished fabric manufacturer, and should be the same as those used during the manufacture of the liferaft.

.6 On each test specimen the bonding between the adhesive or weld and the coating should be initially measured to determine the surface receptivity.

.7 The adhesion of the coating to the base textile is then measured by cutting through one coating layer to initiate the required mode of separation.

.8 After testing in 5.17.13.2.2.3.4 for adhesion of coating to the base textile, the specimen should be immersed for 24 h in a 3% aqueous solution of sodium chloride at 20 ± 2°C. At the end of the immersion period the specimen should be removed from the solution, and while still wet tested by the method specified in ISO 2411.

.2.2.4 *Effects of ageing*

.1 *Folding test*

When tested as prescribed below there should be no cracks, separation of plies or brittleness visible when the samples are inspected under a magnification of 2.

.2 *Tensile test*

When tested as prescribed below the tensile strength after ageing should be not less than 90% of the original tensile strength before ageing.

.3.1 *Ultra-violet resistance*

This test should be performed in accordance with the methods specified in ISO 4892-4:1994 – Open-flame carbon arc lamps, as follows:

.1 Expose the conditioned samples to an enclosed carbon arc lamp without "Corex D" filters for 100 h. The carbons should be Copper Clad Sunshine Arc Type, No. 22 for the upper pair and No. 13 for the lower pair, or equivalent. Only the intended outside surface of the fabric is to be exposed to the arc in the testing apparatus. The specimens should be exposed to water spray, with the apparatus operated so that the specimens are exposed to successive cycles of 102 min of light without spray and 18 min of light with spray. The black panel temperature should be 80 ± 5°C. The total exposure time should be 100 h.

.2 Test the tensile strength of the material after exposure following the procedure in 5.17.13.2.2.1. The tensile strength should be not less than 90% of the original tensile strength before ageing.

.3 The exposed material should be bent, more heavily coated side out, around a 3.2 mm mandrel and examined visually for cracking. There should be no cracking.

.3.2 *Alternative ultra-violet resistance*

Alternatively, this test may be performed in accordance with the methods specified in ISO 4892-2 – Xenon arc type testing. The specimens should be exposed under conditions specified below, using a controlled irradiance water cooled Xenon Arc apparatus for a total exposure time of 150 h.

Exposure conditions	Dark cycle (1 hour)	Light cycle (2 hours)
Automatic irradiance (Filter Q/B)	Nil	0.55 W/m^2 – nm at 340 nm
Black panel temperature	38 ±2°C	70 ±2°C
Dry bulb temperature	38 ±2°C	47 ±2°C
Relative humidity	95 ±5%	50 ±5%
Conditioning water	40 ±4°C	45 ±4°C
Water spray	60 min on front and back of specimen	40 min – Nil 20 min – Front of specimen only 60 min – Nil

Only the intended outside surface of the fabric should be exposed to the arc. The tensile strength of the material should be tested after exposure following the procedure in 5.17.13.2.2.1. The tensile strength should be not less than 90% of the original strength before ageing. The exposed material should be bent, with heavily coated side out, around a 3.2 mm mandrel and each coated face examined visually for cracking. There should be no cracking during this examination.

The performance requirements specified in this subparagraph relate to the behaviour of individual specimens under particular conditions of test. As the spectrum of light from the Carbon Arc differs from that of the Xenon Arc, caution should be exercised in interpreting the test results of both methods.

.4 Three separate specimens should be tested as follows:

.1 dimensional stability;

.2 folding; and

.3 tensile strength.

For .1 and .2, cut from the test sample four specimens at least 100 mm square with the sides closely parallel to the warp and weft threads. Measure the dimensions of two specimens accurately for .1. For .3 cut two sets of specimens as in 5.17.13.2.2.1.

.5 When tested as prescribed below the difference in dimensions of the sample before and after ageing should not differ by more than 2%.

.6 Ageing of specimens test procedure:

.1 Freely suspend one specimen each for 5.17.13.2.2.4.4.1 and 5.17.13.2.2.4.4.2, and one set of specimens for 5.17.13.2.2.4.4.3 in air for seven days at 70 ± 2°C. Suspend the other specimens above water in a loosely closed vessel for seven days at 70 ± 2°C.

.2 Remove the two measured specimens from the ageing oven. After 15 min at room temperature measure the dimensions and report the percentage changes in warp and weft directions.

.3 Remove the other two specimens. After 15 min at room temperature fold the specimens consecutively in two directions parallel to the edges at right angles to each other so as to reduce the exposed area of each specimen to one quarter of its original size. Unfold and refold along the same creases but with each fold reversed in direction. After each folding, press the fold by rubbing fingers and thumb along it: inspect the specimens for cracks, separation of plies, stickiness or brittleness.

.4 For the tensile strength test remove the two sets of specimens from the ageing oven. Dry the wet aged specimens for 1 h in air at 70 ±2°C, and then condition both sets for 24 h. Test in accordance with paragraph 5.17.13.2.2.1.

.2.2.5 *Low temperature flexing*

.1 When tested at a temperature no higher than −50°C by the method prescribed below, there should be no visible cracking of the sample when inspected under a magnification of 2. The test should be independently applied to each face of the coated fabric.

.2 The apparatus, preparation of test specimens and test procedure should be as described in ISO 4675, except that:

.1 when tested at the specified low temperature no specimen should show cracks; and

.2 there should be six test specimens, three cut with the long side closely parallel to the warp and three cut with the long side closely parallel to the weft direction.

.2.2.6 *Flex cracking*

After the specimen has been conditioned by exposing the outer face to a 3% aqueous solution of sodium chloride for seven days at 20 ±2°C, it should be tested as described in ISO 7854. After 200,000 flexings no cracking or delamination should be visible when inspected under a magnification of 2.

.2.2.7 *Porosity*

When tested by the method prescribed below and with a pressure of 27.5 kPa applied and maintained beneath the fabric, there should be no signs of any leakage over a minimum period of 5 min.

.1 Test for porosity

A specimen of the fabric should be prepared and tested in accordance with ISO/TR 6065, paragraph A.2.10.2.

.2.2.8 *Oil resistance*

.1 When tested by the method prescribed below, after exposing the outer surface to oil ASTM No. 1, for 2 h at 20 ± 2°C, there should be no separation of coating from textile and no residual tackiness when two exposed faces are pressed together. The coating should not smear when rubbed with a single pass of the finger.

.2 The test should be carried out not less than 16 h after vulcanisation or curing.

.3 The apparatus, preparation of specimens and test procedure should be in accordance with ISO/TR 6065, paragraph A.2.5. Each coated face should be tested.

.2.2.9 *Weft distortion*

The weft distortion should be not more than the equivalent of 100 mm maximum over a fabric width of 1.5 m. A line should be drawn across the fabric at right angles to the selvedge. The weft distortion, skew and/or bow should be measured.

.2.2.10 *Resistance to blocking*

.1 When tested by the method prescribed below the fabric should exhibit no blocking.

.2 The preparation of specimens and test procedure should be in accordance with ISO 5978 except that the duration of time under load should be seven days.

.2.2.11 *Hydrolysis resistance for thermoplastic coated materials only*

.1 When tested by the methods prescribed below, the following performance values should be achieved:

.1 Coating adhesion	50 N/50 mm minimum
.2 Blocking resistance	100 g maximum
.3 Folding test	No cracks, delamination or visual deterioration

.2 The following test requirements should apply to fabrics or test specimens which have been stored for 12 weeks over water in a closed container at 93°C.

.3 The following test should be performed after drying the specimens for 1 h at 80 ±2°C, and conditioning at 20 ± 2°C, 65% RH for 24 h.

.4 The coating adhesion of the stored material specimen should be made up and tested in accordance with paragraph 5.17.13.2.2.3 after the requirements of 5.17.13.2.2.11.2 above have been carried out.

.5 The blocking resistance should be tested in accordance with paragraph 5.17.13.2.2.10.

.6 Two test samples 100 ±2 mm square should be cut from the stored material. The samples should be folded as defined in section 5.17.13.2.2.4.6.3 and examined for evidence of cracks, ply separation, stickiness or brittleness.

.2.2.12 *Ozone resistance*

.1 When tested by the method prescribed below, no cracks should be visible at a magnification of 5.

.2 The preparation of samples and test procedure should be in accordance with specification ISO 3011.

The following conditions should apply:

.1 Ozone concentration	50 pphm
.2 Temperature	20 ±2°C
.3 Exposure time	8 h
.4 Mandrel diameter	6 × sample thickness.

.2.3 Fabric used for the manufacture of outer canopies should meet the following requirements:

.2.3.1 *Tensile strength*

When tested by the method prescribed in 5.17.13.2.2.1, the tensile strength should be minimum 930 N/50 mm of width in warp and weft directions.

.2.3.2 *Tear strength*

When tested by the method prescribed in 5.17.13.2.2.2, the tear strength should be minimum 490 N in warp and weft direction.

.2.3.3 *Low temperature flexing*

When tested at a temperature not higher than –30°C by the method prescribed in 5.17.13.2.2.5, there should be no visible cracking of the sample when inspected under a magnification of 2.

The test should be independently applied to each face of the coated fabric.

.2.3.4 *Waterproofness*

.1 When tested by the method prescribed below, no water should pass through the cone within 30 min. The coated fabric should not contain any material that is known to be injurious to a survivor drinking rainwater collected from the canopy.

Fabrics may be coated on one or both sides.

.2 The test specimen should be cut to a size of 300 mm × 300 mm and tested in accordance with the following procedure:

Fold the specimen twice at right angles and open it out into the form of a cone. Secure the cone with a paper clip and insert it into a suitable funnel supported on a flask. Pour 500 mℓ of water into the cone. Record any penetration of water to the outside of the cone after 30 min.

.2.3.5 *Surface receptiveness and adhesion of surface coating*

When tested by the method prescribed in 5.17.13.2.2.3, the surface receptiveness on either face should not be less than 25 N/50 mm width surface.

For coating adhesion a minimum of 25 N/50 mm is required.

.2.3.6 *Colour*

The liferaft canopy should be evaluated after the mooring out test in 5.5 or an equivalent method using artificial light to determine whether the coating is sufficiently colour-fast.

.2.3.7 *Effects of ageing*

.1 Folding test

When tested by the method prescribed in 5.17.13.2.2.4 there should be no cracks, separation of plies or brittleness visible when the samples are inspected under a magnification of 2.

.2 Tensile test

When tested by the method prescribed in 5.17.13.2.2.4 at least 90% of the original tensile strength should be retained in both warp and weft direction.

.2.4 Fabric used for the manufacture of inner canopies should meet the following requirements:

.2.4.1 *Tensile strength*

When tested by the method prescribed in 5.17.13.2.2.1 the tensile strength should be minimum 100 N/50 mm of width in warp and weft directions.

2.4.2 *Porosity*

As the inner canopy serves as a barrier to provide a static layer of air, it should either be of a close weave construction or have a low porosity to air.

5.18 Additional tests applicable to automatically self-righting liferafts only

5.18.1 Rigid automatic self-righting liferafts should be tested to the righting test in 5.17.2.1 and 5.17.2.2.

5.18.2 A suitable means should be provided to rotate the liferaft about a longitudinal axis to any angle of heel in calm water and then release it. The liferaft should be fully equipped, with no one on board, with entrances and openings in the as-packed condition and, in the case of an inflatable liferaft, fully inflated. The liferaft should be incrementally rotated to angles of heel up to and including 180° and should be released. After release, the liferaft should always return to the upright position without any assistance. Righting action should be positive and continuous, and complete righting should occur within the time difference between the liferaft reaching boardable shape, as determined by 5.17.3.1, at ambient temperature, and 1 min.

5.19 Submergence test for automatically self-righting and canopied reversible liferafts

The liferaft, if inflatable and in its packed condition, should be submerged to a depth of at least 4 m. A rigid liferaft should be released at this depth, and, if an inflatable liferaft, initiate inflation at this depth. The liferaft should float to the surface and come to its designed operational condition ready to be boarded from the sea in a sea state of at least 2 m significant wave height in association with a wind force of Beaufort force 6.

5.20 Wind velocity tests

5.20.1 The Administration should from a range of liferafts require at least:

.1 one liferaft from a range 6 to 25 persons capacity, provided the material and construction arrangements are similar; and

.2 each liferaft greater than 25 persons capacity, except in the case where it can be shown that the material and construction arrangements deem this unnecessary,

to be tested under the conditions of wind velocity given in the following paragraphs.

5.20.2 The liferaft or liferafts in the packed condition with the entrance so arranged that it will be open on inflation, but without the container, should be inflated in a wind velocity of 30 m/s and should be left in this condition for 10 min.

5.20.3 During the above-mentioned conditions, whenever practicable, the liferaft or liferafts should be swung over approximately 30° to starboard, from that position to approximately 30° to port and return to the starting position.

5.20.4 On completion of these first stage tests, there should be no detachment of the arch support or canopy from the upper buoyancy tube or other damage which affects the efficient function of the liferaft.

5.20.5 Then the liferaft or liferafts should be exposed to the above-mentioned wind velocity for 5 min in each of the following conditions:

.1 with the entrance to the wind open and the other entrances closed, if there is more than one entrance;

.2 with the entrance to the wind closed and the other entrances open, if there is more than one entrance; and

.3 with all entrances closed.

The liferaft or liferafts should show no sign of damage affecting its/their efficient function as a result of this test.

5.21 Test for self-draining of floors of canopied reversible liferafts and automatically self-righting liferafts

5.21.1 Water should be pumped into the interior of the liferaft, while it is afloat, at a rate of 2,300 ℓ per minute for 1 min.

5.21.2 After the water has been shut off and has drained, there should be no appreciable accumulation of water in the liferaft.

5.21.3 If a liferaft is divided into separate areas, by thwarts or other means, each such area should be subjected to the test.

5.22 Liferaft light tests

The liferaft lights should be subjected to the tests prescribed in 10.1.

6 Lifeboats

6.1 Definitions and general conditions

6.1.1 Except as specified otherwise, the mass of an average person as used herein should be taken to be 75 kg for a lifeboat intended for a passenger ship or 82.5 kg for a lifeboat intended for a cargo ship.

6.1.2 When weights are placed in the lifeboat to simulate the effects of an occupant sitting in a seat, the centre of gravity of the weight in each seat should be placed 300 mm above the seat pan along the seat back.

6.2 Lifeboat material tests

Material fire-retardancy test

6.2.1 The hull and canopy material should be flame tested to determine its fire-retarding characteristics by placing a test specimen in a flame. After removal from the flame the burning time and burning distance should be measured and should be to the satisfaction of the Administration.

Lifeboat buoyant material test

6.2.2 When inherent buoyant material is required, the material should be subjected to the tests prescribed in 2.7 except that in 2.7.6.3 high octane petroleum spirit should be substituted for diesel oil.

6.2.3 In addition to the test in 6.2.2, specimens of the material should be immersed in each of the following for a period of 14 days under a 100 mm head:

- **.1** two specimens in crude oil;
- **.2** two specimens in marine fuel oil (grade C);
- **.3** two specimens in diesel oil (grade A);
- **.4** two specimens in high octane petroleum spirit; and
- **.5** two specimens in kerosene.

6.2.4 The specimens should be tested as supplied by the manufacturer and at normal room temperature (approximately 18°C).

6.2.5 Two additional specimens, which have already been subjected to the temperature cycling tests, should be tested against high octane petroleum spirit and afterwards subjected to the water absorption test as prescribed in 2.7.5 to 2.7.8.

6.2.6 The dimensions of the specimens should be recorded at the beginning and end of these tests.

6.2.7 The reduction of buoyancy must not exceed 5% and the specimens should show no sign of damage such as shrinking, cracking, swelling, dissolution or change of mechanical qualities.

6.3 Lifeboat overload test

Davit-launched lifeboats

6.3.1 The unloaded lifeboat should be placed on blocks or suspended from the lifting hooks and sights should be erected for measuring keel sag. The measurements required in 6.3.4 should then be made.

6.3.2 The lifeboat should then be loaded with properly distributed weights to represent the fully equipped lifeboat loaded with the full complement of persons for the type of ship for which it is to be approved. The measurements required in 6.3.4 should again be made.

6.3.3 Additional weights should then be added so that the suspended load is 25%, 50%, 75% and 100% greater than the weight of the fully equipped and loaded lifeboat. In the case of metal lifeboats, the testing should stop at 25% overload. The weights for the various overload conditions should be distributed in proportion to the loading of the lifeboat in its service condition, but the weights used to represent the persons need not be placed 300 mm above the seat pan. Testing by filling the lifeboat with water should not be accepted as this method of loading does not give the proper distribution of weight. Machinery may be removed in order to avoid damage to it, in which case weights should be added to the lifeboat to compensate for the removal of such machinery. At each incremental overload, the measurements required in 6.3.4 should be made.

6.3.4 The following should be measured and recorded at each condition of load specified in 6.3.1 through 6.3.3:

.1 deflection of keel amidships;

.2 change in length as measured between the top of stem and stern posts;

.3 change in breadth over the gunwale at the quarter length forward, amidships and the quarter length aft; and

.4 change in depth measured from gunwale to keel.

6.3.5 The keel deflection and change in breadth in 6.3.4.1 and 6.3.4.3 should not exceed 1/400 of the lifeboat's length when the lifeboat is subjected to 25% overload; the results at 100% overload, if required by 6.3.3, should be approximately in proportion to those obtained at 25% overload.

6.3.6 The weights should then be removed and the dimensions of the lifeboat checked. No significant residual deflection should result. Any permanent deflection as a result of these tests should be recorded. If the lifeboat is made of GRP, such measurement should be taken after a lapse of time sufficient to permit the GRP to recover its original form (approximately 18 h).

Free-fall lifeboats

6.3.7 It should be demonstrated that the lifeboat has sufficient strength to withstand the forces acting upon it when loaded with a distributed mass equal to the mass of the number of persons for which it is to be approved and its equipment when free-fall launched from a height of 1.3 times the height for which it is to be approved. If the lifeboat is normally ramp-launched, and a ramp is not available, this test may be conducted by dropping the lifeboat vertically with the keel at the same angle that normally occurs during water entry.

6.3.8 After this test the lifeboat should be unloaded, cleaned and carefully examined to detect the position and extent of damage that may have occurred as a result of this test. An operational test should then be conducted in accordance with 6.10.1. After this test the lifeboat should again be unloaded, cleaned, and inspected for possible damage.

6.3.9 This test should be considered successful if the lifeboat passes the operational test to the satisfaction of the Administration; no damage has been sustained that would affect the lifeboat's efficient functioning; and any deflections of the hull or canopy as measured during the test would not cause injury to lifeboat occupants.

6.4 Davit-launched lifeboat impact and drop test

Impact test

6.4.1 The fully equipped lifeboat, including its engine, should be loaded with weights equal to the mass of the number of persons for which the lifeboat is to be approved. In totally enclosed lifeboats, representative safety belts and fastenings which will experience high loads as a result of the impact should be secured about weights equal to 100 kg to simulate holding a person during the test. The weights should be distributed to represent the normal loading in the lifeboat. Skates or fenders, if required, should be in position. The lifeboat, in a free hanging position, should be pulled laterally to a position so that when released it will strike a fixed rigid vertical surface at a velocity of 3.5 m/s. It should be released to impact against the rigid vertical surface.

6.4.2 In the case of totally enclosed lifeboats, the acceleration forces should be measured and evaluated in accordance with 6.17 at different positions within the prototype lifeboat to determine the most severe occupant exposure to acceleration considering the effects of fenders, lifeboat elasticity, and seating arrangement.

Drop test

6.4.3 The fully equipped lifeboat, with its engine, should be loaded with weights equal to the mass of the maximum number of persons for which the lifeboat is to be approved. Included in this loading should be a weight of 100 kg loaded in one of each type of seat installed in the lifeboat. The remainder of the weights should be distributed to represent the normal loading condition but need not be placed 300 mm above the seatpan. The lifeboat should then be suspended above the water so that the distance from the lowest point of the lifeboat to the water is 3 m. The lifeboat should then be released so that it falls freely into the water.

6.4.4 The drop test should be conducted with the lifeboat that was used in the impact test.

Operational test after impact and drop test

6.4.5 After the impact and drop tests, the lifeboat should be unloaded, cleaned and carefully examined to detect the position and extent of damage that may have occurred as a result of the tests. An operational test should then be conducted in accordance with 6.10.1.

Acceptability criteria for impact and drop tests

6.4.6 After the tests required in this section, the lifeboat should be unloaded, cleaned, and inspected for possible damage.

6.4.7 The impact and drop tests should be considered successful if:

.1 no damage has been sustained that would affect the lifeboat's efficient functioning;

.2 the damage caused by the impact and drop tests has not increased significantly as a result of the test specified in 6.4.5;

.3 machinery and other equipment has operated to full satisfaction;

.4 no significant ingress of seawater has occurred; and

.5 accelerations measured during the impact and subsequent rebound, if required during the impact test, are in compliance with the criteria of either 6.17.9 to 6.17.12 or 6.17.13 to 6.17.17 when using the emergency limits specified in table 2 or table 3, respectively.

6.5 Free-fall lifeboat free-fall test

Required free-fall tests

6.5.1 A lifeboat designed for free-fall launching should be subjected to test launches conducted from the height at which the lifeboat is intended to be stowed taking into account conditions of unfavourable list and trim, unfavourable locations of the centre of gravity, and extreme conditions of load.

6.5.2 During the free-fall launches required in this section, acceleration forces should be measured and the data evaluated in accordance with 6.17 at different locations in the lifeboat to determine the worst occupant exposure to acceleration taking into consideration the seating arrangement.

6.5.3 The tests required in this section may be conducted with correctly scaled models that are at least 1 m in length. As a minimum, the dimensions and mass of the lifeboat, the location of its centre of gravity, and its second moment of mass, must be scaled in a reasonable manner. Depending on the construction and behaviour of the free-fall lifeboat, other parameters may also have to be reasonably scaled to effect correct behaviour of the model. If models are used, sufficient full-scale tests should be conducted to verify the accuracy of the model measurements. As a minimum, the following full-scale tests should be conducted with the ship on an even keel using the same type of launching arrangement as the production lifeboat and from the height for which the lifeboat is to be approved:

- **.1** lifeboat fully loaded;
- **.2** lifeboat loaded with its required equipment and minimum launching crew only;
- **.3** lifeboat loaded with its required equipment and one half of the full complement of persons distributed in the forward half of the seating positions of the lifeboat; and
- **.4** lifeboat loaded with its required equipment and one half of the full complement of persons seated in the after half of the seating positions of the lifeboat.

Acceptability criteria for free-fall tests

6.5.4 The free-fall tests required in this section should be considered acceptable if:

- **.1** the acceleration forces are in compliance with the "Training" condition specified in tables 2 and 3 of 6.17 during the launch, free-fall, and subsequent water entry for those tests with the ship on even keel;

.2 the acceleration forces are in compliance with the "Emergency" condition specified in tables 2 and 3 of 6.17 during the launch, free-fall, and subsequent water entry for those tests with the ship under unfavourable conditions of list and trim; and

.3 the lifeboat makes positive headway immediately after water entry.

6.6 Lifeboat seating strength test

Davit-launched lifeboats

6.6.1 The seating should be loaded with a mass of 100 kg in each position allocated for a person to sit in the lifeboat. The seating should be able to support this loading without any permanent deformation or damage.

Free-fall lifeboats

6.6.2 The seats experiencing the highest acceleration forces, and those seats which are supported in a manner different from the other seats in the lifeboat, should be loaded with a mass of 100 kg. The load should be arranged in the seat so that both the seat back and the seat pan are affected. The seating should be able to support this load during a free-fall launch from a height of 1.3 times the approved height, without any permanent deformation or damage. This test may be conducted as part of the test in 6.3.7 to 6.3.9.

6.7 Lifeboat seating space test

6.7.1 The lifeboat should be fitted with its engine and its equipment. The number of persons for which the lifeboat is to be approved, having an average mass of 75 kg for a lifeboat intended for a passenger ship or 82.5 kg for a lifeboat intended for a cargo ship and wearing a lifejacket and any other essential equipment, should be able to board the lifeboat and be properly seated within a period of 3 min in the case of a lifeboat intended for a cargo ship and as rapidly as possible in the case of a lifeboat intended for a passenger ship. The lifeboat should then be manoeuvred and all equipment on board tested by an individual to demonstrate that the equipment can be operated without difficulty and without interference with the occupants.

6.7.2 The surfaces on which persons might walk should be visually examined to determine that they have a non-skid finish.

6.8 Lifeboat freeboard and stability tests

Flooded stability test

6.8.1 The lifeboat should be loaded with its equipment. If provision lockers, water tanks and fuel tanks cannot be removed, they should be flooded or filled to the final waterline resulting from the test in 6.8.3. Lifeboats fitted with watertight stowage compartments to accommodate individual drinking water containers should have these containers aboard and placed in the stowage compartments which should be sealed watertight during the flooding tests. Ballast of equivalent weight and density should be substituted for the engine and any other installed equipment that can be damaged by water.

6.8.2 Weights representing persons who would be in the water when the lifeboat is flooded (water level more than 500 mm above the seat pan) may be omitted. Weights representing persons who would not be in the water when the lifeboat is flooded (water level less than 500 mm above seat pan) should be placed in the normal seating positions of such persons with their centre of gravity approximately 300 mm above the seat pan. Weights representing persons who would be partly submerged in the water when the lifeboat is flooded (water level between 0 and 500 mm above the seat pan) should additionally have an approximate density of 1 kg/dm^3 (for example water ballast containers) to represent a volume similar to a human body.

6.8.3 When loaded as specified in 6.8.1 and 6.8.2, the lifeboat should have positive stability when filled with water to represent flooding which would occur when the lifeboat is holed in any one location below the water-line assuming no loss of buoyancy material and no other damage. Several tests may have to be conducted if holes in different areas would create different flooding conditions.

Freeboard test

6.8.4 The lifeboat with its engine should be loaded with a mass equal to that of all the equipment. One half of the number of persons for which the lifeboat is to be approved should be seated in a proper seating position on one side of the centreline. The freeboard should then be measured on the low side.

6.8.5 This test should be considered successful if the measured freeboard on the low side is not less than 1.5% of the lifeboat's length or 100 mm, whichever is greater.

6.9 Release mechanism test

Davit-launched lifeboats

6.9.1 The lifeboat with its engine fitted should be suspended from the release mechanism just clear of the ground or the water. The lifeboat should be loaded so that the total mass equals 1.1 times the mass of the lifeboat, all its equipment and the number of persons for which the lifeboat is to be approved. The lifeboat should be released simultaneously from each fall to which it is connected without binding or damage to any part of the lifeboat or the release mechanism.

6.9.2 It should be confirmed that the lifeboat will simultaneously release from each fall to which it is connected when fully waterborne in the light condition and in a 10% overload condition.

6.9.3 The release mechanism should be mounted on a tensile strength testing device. The load should be increased to at least six times the working load of the release mechanism without failure of the release mechanism.

6.9.4 It should be demonstrated that the release mechanism can release the fully equipped lifeboat when loaded with weights equal to the mass of the number of persons for which the lifeboat is to be approved, when the lifeboat is being towed at speeds up to 5 knots. In lieu of a waterborne test, this test may be conducted as follows:

- **.1** a force equal to 25% of the safe working load of the hook should be applied to the hook in the lengthwise direction of the boat at an angle of 45° to the vertical. This test should be conducted in the aftward as well as the forward direction;
- **.2** a force equal to the safe working load of the hook should be applied to the hook in an athwartships direction at an angle of 20° to the vertical. This test should be conducted on both sides;
- **.3** a force equal to the safe working load of the hook should be applied to the hook in a direction half-way between the positions of tests 1 and 2 (i.e., 45° to the longitudinal axis of the boat in plan view) at an angle of 33° to the vertical. This test should be conducted in four positions.

There should be no damage to the hook as a result of this test, and in the case of a waterborne test, there should be no damage to the lifeboat or its equipment.

Free-fall lifeboats

6.9.5 It should be demonstrated that the free-fall release mechanism can operate effectively when loaded with a force equal to at least 200% of the normal load caused by the fully equipped lifeboat when loaded with the number of persons for which it is to be approved.

6.9.6 The release mechanism should be mounted on a tensile strength testing device. The load should be increased to at least six times the working load of the release mechanism without failure of the release mechanism.

6.10 Lifeboat operational test

Operation of engine and fuel consumption test

6.10.1 The lifeboat should be loaded with weights equal to the mass of its equipment and the number of persons for which the lifeboat is to be approved. The engine should be started and the lifeboat manoeuvred for a period of at least 4 h to demonstrate satisfactory operation. The lifeboat should be run at a speed of not less than 6 knots for a period which is sufficient to ascertain the fuel consumption and to establish that the fuel tank has the required capacity. The maximum towing force of the lifeboat should be determined. This information should be used to determine the largest fully loaded liferaft the lifeboat can tow at 2 knots. The fitting designated for towing other craft should be secured to a stationary object by a tow rope. The engine should be operated ahead at full speed for a period of at least 2 min, and the towing force measured and recorded. There should be no damage to the towing fitting or its supporting structure. The maximum towing force of the lifeboat should be recorded on the type approval certificate.

Cold engine starting test

6.10.2 The engine may be removed from the lifeboat for this test; however, it should be equipped with accessories and the transmission that will be used in the lifeboat. The engine, along with its fuel and coolant, should be placed in a chamber at a temperature of –15°C.

6.10.3 The temperature of the fuel, lubricating oil and cooling fluid (if any) should be measured at the beginning of this test and should not be higher than –15°C. Samples of each fluid at this temperature should be collected in a container for observation.

6.10.4 The engine should be started three times. The first two times, the engine should be allowed to operate long enough to demonstrate that it runs at operating speed. After the first two starts, the engine should be allowed to stand until all parts have again reached chamber temperature. After the third start, the engine should be allowed to continue to run for at least 10 min and during this period the transmission should be operated through its gear positions.

Engine-out-of-water test

6.10.5 The engine should be operated for at least 5 min at idling speed under conditions simulating normal storage. The engine should not be damaged as a result of this test.

Submerged engine test

6.10.6 The engine should be operated for at least 5 min while submerged in water to the level of the centreline of the crankshaft with the engine in a horizontal position. The engine should not be damaged as a result of this test.

Compass

6.10.7 It should be determined that the compass performance is satisfactory and that it is not unduly affected by magnetic fittings and equipment in the lifeboat.

Survival recovery test

6.10.8 It should be demonstrated by test that it is possible to bring helpless people on board the lifeboat from the sea.

6.11 Lifeboat towing and painter release test

Towing test

6.11.1 It should be demonstrated that the fully equipped lifeboat, loaded with a properly distributed mass equal to the mass of the number of persons for which it is to be approved, can be towed at a speed of not less than 5 knots in calm water and on an even keel. There should be no damage to the lifeboat or its equipment as a result of this test.

Davit-launched lifeboat painter release test

6.11.2 It should be demonstrated that the painter release mechanism can release the painter on a fully equipped and loaded lifeboat that is being towed at a speed of not less than 5 knots in calm water.

6.11.3 The painter release mechanism should be tested in several distinct directions of the upper hemisphere not obstructed by the canopy or other constructions in the lifeboat. The directions specified in 6.9.4 should be used if possible.

6.12 Lifeboat light tests

The lifeboat light should be subjected to the tests prescribed in 10.1.

6.13 Canopy erection test

6.13.1 This test is required only for partially enclosed lifeboats. During the test the lifeboat should be loaded with the number of persons for which it is to be approved.

6.13.2 If the lifeboat is partially enclosed it should be demonstrated that the canopy can be easily erected by not more than two persons.

6.14 Additional tests for totally enclosed lifeboats

Self-righting test

6.14.1 A suitable means should be provided to rotate the lifeboat about a longitudinal axis to any angle of heel and then release it. The lifeboat, in the enclosed condition, should be incrementally rotated to angles of heel up to and including 180° and should be released. After release, the lifeboat should always return to the upright position without the assistance of the occupants. These tests should be conducted in the following conditions of load:

- **.1** when the lifeboat with its engine is loaded in the normal position with properly secured weights representing the fully equipped lifeboat with a full complement of persons on board. The weight used to represent each person, assumed to have an average mass of 75 kg, should be secured at each seat location and have its centre of gravity approximately 300 mm above the seatpan so as to have the same effect on stability as when the lifeboat is loaded with the number of persons for which it is to be approved; and
- **.2** when the lifeboat is in the light condition.

6.14.2 At the beginning of these tests, the engine should be running in neutral position and:

.1 unless arranged to stop automatically when inverted, the engine should continue to run when inverted and for 30 min after the lifeboat has returned to the upright position;

.2 if the engine is arranged to stop automatically when inverted, it should be easily restarted and run for 30 min after the lifeboat has returned to the upright position.

Flooded capsizing test

6.14.3 The lifeboat should be placed in the water and fully flooded until the lifeboat can contain no additional water. All entrances and openings should be secured to remain open during the test.

6.14.4 Using a suitable means, the lifeboat should be rotated about a longitudinal axis to a heel angle of 180° and then released. After release, the lifeboat should attain a position that provides an above-water escape for the occupants.

6.14.5 For the purpose of this test, the mass and distribution of the occupants may be disregarded. However, the equipment, or equivalent mass, should be secured in the lifeboat in the normal operating position.

Engine inversion test

6.14.6 The engine and its fuel tank should be mounted on a frame that is arranged to rotate about an axis equivalent to the longitudinal axis of the boat. A pan should be located under the engine to collect any oil which may leak from the engine so that the quantity of such oil can be measured.

6.14.7 The following procedure should be followed during this test:

.1 start the engine and run it at full speed for 5 min;

.2 stop the engine and rotate it in a clockwise direction through 360°;

.3 restart the engine and run it at full speed for 10 min;

.4 stop the engine and rotate it in a counter-clockwise direction through 360°;

.5 restart the engine, run it at full speed for 10 min, and then stop the engine;

.6 allow the engine to cool;

.7 restart the engine and run it at full speed for 5 min;

.8 rotate the running engine in a clockwise direction through 180°, hold at the 180° position for 10 s, and then rotate it 180° further in a clockwise direction to complete one revolution;

.9 if the engine is arranged to stop automatically when inverted, restart it;

.10 allow the engine to continue to run at full speed for 10 min;

.11 shut the engine down and allow it to cool;

.12 repeat the procedure in 6.14.7.7 through 6.14.7.11, except that the engine should be turned in a counter-clockwise direction;

.13 restart the engine and run it at full speed for 5 min;

.14 rotate the engine in a clockwise direction through 180° and stop the engine. Rotate it 180° further to complete a full clockwise revolution;

.15 restart the engine wand run it at full speed for 10 min;

.16 repeat the procedure in 6.14.7.14, turning the engine counter-clockwise;

.17 restart the engine, run it at full speed for 10 min and then shut it down; and

.18 dismantle the engine for examination.

6.14.8 During these tests, the engine should not overheat, fail to operate, or leak more than 250 mℓ of oil during any one inversion. When examined after being dismantled the engine should show no evidence of overheating or excessive wear.

6.15 Air supply test for lifeboats with a self-contained air support system

All entrances and openings of the lifeboat should be closed and the air supply to the inside of the lifeboat turned on to the design air pressure. The engine should then be run at revolutions necessary to achieve full speed with the fully loaded boat including all persons and with the sprinkler system in use for a period of 5 min, stopped for 30 s, then restarted for a total running time of 10 min. During this time the atmospheric pressure within the enclosure should be continuously monitored to ascertain that a small positive air pressure is maintained within the lifeboat and to confirm that noxious gases cannot enter. The internal air pressure should never fall below the outside atmospheric pressure, nor should it exceed outside atmospheric pressure by more than 20 hPa during the test. It should be ascertained, by starting the engine with air supply turned off, that when the air supply is depleted, automatic means are activated to prevent a dangerous underpressure of more than 20 hPa being developed within the lifeboat.

6.16 Additional tests for fire-protected lifeboats

Fire test

6.16.1 The lifeboat should be moored in the centre of an area which is not less than five times the maximum projected plan area of the lifeboat. Sufficient kerosene should be floated on the water within the area so that when ignited it will sustain a fire which completely envelops the lifeboat for the period of time specified in 6.16.3. The boundary of the area should be capable of completely retaining the fuel.

6.16.2 The engine should be run at full speed; however, the propeller need not be turning. The gas- and fire-protective systems should be in operation throughout the fire test.

6.16.3 The kerosene should be ignited. It should continue to burn and envelop the lifeboat for 8 min.

6.16.4 During the fire test, the temperature should be measured and recorded as a minimum at the following locations:

.1 at not less than 10 positions on the inside surface of the lifeboat;

.2 at not less than five positions inside the lifeboat at locations normally taken by occupants and away from the inside surface; and

.3 on the external surface of the lifeboat.

The positions of such temperature recorders should be to the satisfaction of the Administration. The method of temperature measurement should allow the maximum temperature to be recorded.

6.16.5 The atmosphere inside the lifeboat should be continuously sampled and representative retained samples should be analysed for the presence and quantity of essential, toxic, and injurious gases or substances. The analysis should cover the range of anticipated gases or substances that may be produced and which can vary according to the materials and fabrication techniques used to manufacture the lifeboat. The analysis should indicate that there is sufficient oxygen and no dangerous levels of toxic or injurious gases or substances.

6.16.6 The pressure inside the lifeboat should be continuously recorded to confirm that a positive pressure is being maintained inside the lifeboat.

6.16.7 At the conclusion of the fire test, the condition of the lifeboat should be such that it could continue to be used in the fully-loaded condition.

Note: The Administration may waive this test for any totally enclosed lifeboat which is identical in construction to another lifeboat which has successfully completed this test, provided the lifeboat differs only in size, and retains essentially the same form. The protective system should be as effective as that of the lifeboat tested. The water delivery rate and film thickness at various locations around the hull and canopy should be equal to or exceed the measurements made on the lifeboat originally fire tested.

Water spray tests

6.16.8 Start the engine and the spray pump. With the engine running at its designed output, the following should be measured to obtain the rated value and speed:

.1 the rpm of the engine and the pump to obtain the rated speed;

.2 the pressure at the suction and delivery side of the pump to obtain the rated water pressure.

6.16.9 With the lifeboat in an upright position, on an even keel and in the light condition, run the pump at the rated speed. Measure the delivery rate of water or the thickness of the sprayed water film at the external surface of the lifeboat. The delivery rate of water or the sprayed water film thickness over the lifeboat should be to the satisfaction of the Administration.

6.16.10 Successively trim the lifeboat 5° by the head and 5° by the stern, and heel it 5° to port and 5° to starboard. In each condition the sprayed water film should cover the whole surface of the lifeboat.

6.17 Measuring and evaluating acceleration forces

Selection, placement and mounting of accelerometers

6.17.1 The accelerometers used to measure the acceleration forces in the lifeboat should:

.1 have adequate frequency response for the test in which they are to be used but the frequency response should at least be in the range of 0 to 200 Hz;

.2 have adequate capacity for the acceleration forces that will occur during the tests;

.3 have an accuracy of ±5%.

6.17.2 Accelerometers should be placed in the lifeboat, parallel to the principal axes of the lifeboat, at those locations necessary to determine the worst occupant exposure to acceleration.

6.17.3 The accelerometers should be mounted on a rigid part of the interior of the lifeboat in a manner to minimize vibration and slipping.

6.17.4 A sufficient number of accelerometers should be used at each location at which acceleration forces are measured so that all likely acceleration forces at that location can be measured.

6.17.5 The selection, placement, and mounting of the accelerometers should be to the satisfaction of the Administration.

Recording method and rate

6.17.6 The measured acceleration forces may be recorded on magnetic media as either an analog or a digital signal or a paper plot of the acceleration signal may be produced.

6.17.7 If the acceleration forces are to be recorded and stored as a digital signal, the sampling rate should be at least 500 samples per second.

6.17.8 Whenever an analog acceleration signal is converted to a digital signal, the sampling rate should be at least 500 samples per second.

Evaluation with the dynamic response model

6.17.9 The dynamic response model is the preferred method to evaluate potential for the occupant in a lifeboat to be injured by exposure to acceleration forces. In the dynamic response model, the human body is idealized as a single-degree-of-freedom spring-mass acting in each co-ordinate direction as shown in figure 3. The response of the body mass relative to the seat support, which is excited by the measured accelerations, can be evaluated using a procedure acceptable to the Administration. The parameters to be used in the analysis are shown in table 1 for each co-ordinate direction.

6.17.10 Before performing the dynamic response analysis, the measured accelerations should be oriented to the primary axes of the seat.

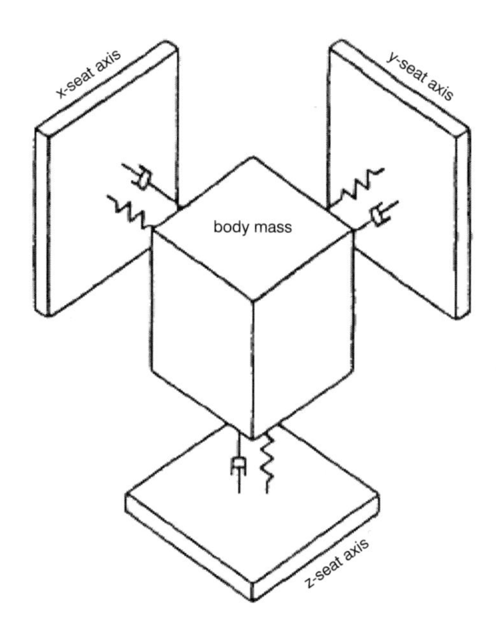

Figure 3 – *Independent single-degree-of-freedom representation of human body*

Table 1 – *Parameters of the dynamic response model*

Co-ordinate axis	Natural frequency (rad/s)	Damping ratio
X	62.8	0.100
Y	58.0	0.090
Z	52.9	0.224

6.17.11 The desired outcome from the dynamic response analysis is the displacement time-history of the body mass relative to the seat support in each co-ordinate direction.

6.17.12 At all times, the following expression should be satisfied:

$$\sqrt{\left(\frac{d_x}{S_x}\right)^2 + \left(\frac{d_y}{S_y}\right)^2 + \left(\frac{d_z}{S_z}\right)^2} \leq 1$$

where d_x, d_y and d_z are the concurrent relative displacements of body mass with respect to the seat support, in the *x*, *y* and *z* body axes, as computed from the dynamic response analysis and S_x, S_y and S_z are relative displacements which are presented in table 2 for the appropriate launch condition.

Table 2 – *Suggested displacement limits for lifeboats*

Acceleration direction	Displacement (cm)	
	Training	Emergency
+ *X* - - Eyeballs in	6.96	8.71
– *X* - - Eyeballs out	6.96	8.71
+ *Y* - - Eyeballs right	4.09	4.95
– *Y* - - Eyeballs left	4.09	4.95
+ Z - - Eyeballs down	5.33	6.33
– Z - - Eyeballs up	3.15	4.22

Evaluation using the SRSS method

6.17.13 In lieu of the procedure in 6.17.9 to 6.17.12, the potential for an occupant in a lifeboat to become injured by an acceleration can be evaluated using the procedure presented in this section.

6.17.14 Before performing the SRSS analysis, the measured accelerations should be oriented to the primary axes of the seat.

6.17.15 Full-scale acceleration data should be filtered with no less than the equivalent of a 20 Hz low-pass filter. Any filtering procedure acceptable to the Administration may be used.

6.17.16 Acceleration data measured on a model should be filtered with a low pass filter having a frequency not less than that obtained with the following expression:

$$f_{\text{model}} = \frac{20}{\sqrt{\frac{L_{\text{model}}}{L_{\text{prototype}}}}}$$

where f_{model} is the frequency of the filter to be used, L_{model} is the length of the model lifeboat, and $L_{\text{prototype}}$ is the length of the prototype lifeboat.

6.17.17 At all times, the following expression should be satisfied:

$$\sqrt{\left(\frac{g_x}{G_x}\right)^2 + \left(\frac{g_y}{G_y}\right)^2 + \left(\frac{g_z}{G_z}\right)^2} \leq 1$$

where g_x, g_y and g_z are the concurrent accelerations in the *x*, *y* and *z* seat axes and G_x, G_y and G_z are allowable accelerations which are presented in table 3 for the appropriate launch condition.

Table 3 – *SRSS acceleration limits for lifeboats*

Acceleration direction	Acceleration	
	Training	Emergency
+ *X* - - Eyeballs in	15.0	18.0
– *X* - - Eyeballs out	15.0	18.0
+ *Y* - - Eyeballs right	7.0	7.0
– *Y* - - Eyeballs left	7.0	7.0
+ *Z* - - Eyeballs down	7.0	7.0
– *Z* - - Eyeballs up	7.0	7.0

7 Rescue boats and fast rescue boats

7.1 Rigid rescue boats

7.1.1 Rigid rescue boats should be subjected to the tests prescribed in 6.2 to 6.12 (except 6.3, 6.4.2, 6.5, 6.6.2, 6.7.1, 6.9.5, 6.9.6, 6.10.1) and 7.2.4.2.

Towing test

7.1.2 The maximum towing force of the rescue boat should be determined. This information should be used to determine the largest fully loaded liferaft the rescue boat can tow at two knots. The fitting designated for towing other craft should be secured to a stationary object by a tow rope. The engine should be operated ahead at full speed for a period of at least 2 min, and the towing force measured and recorded. There should be no damage to the towing fitting or its supporting structure. The maximum towing force of the rescue boat should be recorded on the type approval certificate.

Rigid rescue boat seating test

7.1.3 The rigid rescue boat should be fitted with its engine and all its equipment. The number of persons for which the rescue boat is to be approved, having an average mass of at least 82.5 kg and all wearing lifejackets and immersion suits and any other essential equipment required should then board; one person should lie down on a stretcher of similar dimensions to those shown in figure 4 and the others should be properly seated in the rescue boat. The rigid rescue boat should then be manoeuvred and all equipment on board tested to demonstrate that it can be operated without difficulty or interference with the occupants.

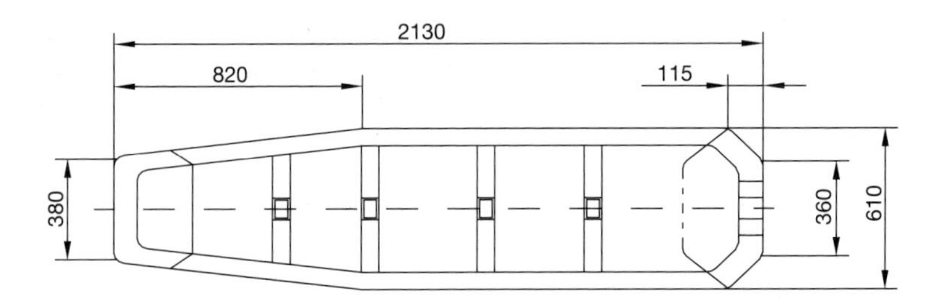

Figure 4 – *Stretcher dimensions (mm)*

Overload test

7.1.4 The boat should be loaded with properly distributed load of four times the weights to represent the equipment and full complement of persons each weighing 82.5 kg for which it is to be approved and suspended for five minutes from its bridle or hooks. The weights should be distributed in proportion to the loading of the boat in its service condition, but the weights used to represent the persons need not be placed 300 mm above the seat pan. The boat and bridle or hooks and fastening device should be examined after the test has been conducted and should not show any signs of damage. Testing by filling the boat with water should not be accepted. This method of loading does not give the proper distribution of weight. Machinery may be removed in order to avoid damage, in which case weights should be added to the boat to compensate for the removal of such machinery.

Operation tests

7.1.5 ***Operation of engine and fuel consumption test***

The boat should be loaded with weights equal to the mass of its equipment and the number of persons for which the boat is to be approved. The engine should be started and the boat manoeuvred for a period of at least 4 h to demonstrate satisfactory operation.

The boat should be run at a speed of not less than 6 knots for a period which is sufficient to ascertain the fuel consumption and to establish that the fuel tank has the required capacity.

7.1.6 Speed and manoeuvring trials should be carried out with engines of various powers to assess the rescue boat's performance (if a rigid rescue boat is equipped with an outboard motor).

Righting test

7.1.7 It should be demonstrated that both with and without engine and fuel or an equivalent mass in place of the engine and fuel tank, the rescue boat is capable of being righted by not more than two persons if it is inverted on the water. In the case of fast rescue boats which are not self-righting, the engine should be running in neutral position and, after stopping automatically or by the helmsman's emergency release switch when inverted, it should be easily restarted and run for 30 min after the rescue boat has returned to the upright position. For rescue boats with inboard engines, the test without engine and fuel is not applicable.

Manoeuvrability test

7.1.8 It should be demonstrated that the rigid rescue boat can be propelled and manoeuvred by its oars or paddles in calm water conditions at a speed of at least 0.5 knots over a distance of at least 25 m, when laden with the number of persons, all wearing lifejackets and immersion suits, for which it is to be approved.

Detailed inspection

7.1.9 The rigid rescue boat, complete in all respects, should be subjected to detailed inspection to ensure that all the requirements are fulfilled.

7.2 Inflated rescue boats

7.2.1 The inflated rescue boats should be subjected to the tests prescribed in 6.4.1, 6.6.1, 6.7.2, 6.9.1 to 6.9.4, 6.10 (except 6.10.1), 6.11, 6.12, 7.1.2, 7.1.3 and 7.1.5 to 7.1.8.

Drop tests

7.2.2 The inflated rescue boat complete with all its equipment and with a mass equivalent to its engine and fuel in the position of its engine and fuel tank should be dropped three times from a height of at least 3 m onto water. The drops should be from the 45° bow-down, level-trim and 45° stern-down attitudes.

7.2.3 On completion of these drop tests the rescue boat and its equipment should be carefully examined and show no signs of damage which would affect their efficient functioning.

Loading tests

7.2.4 The freeboard of the inflated rescue boat should be taken in the various loading conditions as follows:

.1 rescue boat with all its equipment;

.2 rescue boat with all its equipment, engine and fuel, or an equivalent mass positioned to represent engine and fuel;

.3 rescue boat with all its equipment and the number of persons for which it is to be approved having an average mass of 82.5 kg so arranged that a uniform freeboard is achieved at the side buoyancy tubes; and

.4 rescue boat with the number of persons for which it is to be approved and all its equipment, engine and fuel or an equivalent mass to represent engine and fuel and the rescue boat being retrimmed as necessary.

7.2.5 With the rescue boat in any of the conditions prescribed in 7.2.4, the minimum freeboard should be not less than 300 mm at the buoyancy tubes and not less than 250 mm from the lowest part of the transom.

Stability test

7.2.6 The following tests should be carried out with engine and fuel or an equivalent mass in place of the engine and fuel tanks:

.1 the number of persons for which the inflated rescue boat is to be approved should be crowded to one side with half this complement seated on the buoyancy tube, and then to one end. In each case the freeboard should be recorded. Under these conditions the freeboard should be everywhere positive; and

.2 the stability of the rescue boat during boarding should be ascertained by two persons in the rescue boat demonstrating that they can readily assist from the water a third person who is required to feign unconsciousness. The third person should have his back towards the side of the rescue boat so that he cannot assist the rescuers. All persons should wear approved lifejackets.

7.2.7 These stability tests may be carried out with the rescue boat floating in still water.

Damage test

7.2.8 The following tests should be carried out with the inflated rescue boat loaded with the number of persons for which it is to be approved both with and without engine and fuel or an equivalent mass in the position of the engine and fuel tank:

.1 with forward buoyancy compartment deflated;

.2 with the entire buoyancy on one side of the rescue boat deflated; and

.3 with the entire buoyancy on one side and the bow compartment deflated.

7.2.9 In each of the conditions prescribed by 7.2.8. the full number of persons for which the rescue boat is to be approved should be supported within the rescue boat.

Simulated heavy weather test

7.2.10 To simulate use in heavy weather the inflated rescue boat should be fitted with a larger powered engine than is intended to be fitted and driven hard in a wind of force 4 or 5 or equivalent rough water for at least 30 min. As a result of this test the rescue boat should not show undue flexing or permanent strain nor have lost more than minimal pressure.

Swamp test

7.2.11 It should be demonstrated that the rescue boat, when fully swamped, is capable of supporting its full equipment, the number of persons each weighing 82.5 kg for which it is to be approved and a mass equivalent to its engine and full tank. It should also be demonstrated that the rescue boat does not seriously deform in this condition.

Overload tests

7.2.12 The inflated rescue boat should be loaded with four times the mass of the full complement of persons and equipment for which it is to be approved and suspended for five minutes from its bridle at an ambient temperature of +20 ± 3°C with all relief valves inoperative. The rescue boat and bridle should be examined after the test is conducted and should not show any signs of damage.

7.2.13 The inflated rescue boat after 6 h conditioning at a temperature of −30°C should be loaded with 1.1 times the mass of the full complement of persons and equipment for which it is to be approved and suspended for five minutes from its bridle with all relief valves operative. The rescue boat and bridle should be examined after the test is conducted and should not show any signs of damage.

Material tests

7.2.14 The material used in the construction of inflated rescue boats should be tested for the following characteristics and comply with the requirements of an international standard acceptable to the Organization:*

- **.1** tensile strength;
- **.2** tear strength;
- **.3** heat resistance;
- **.4** cold resistance;
- **.5** heat ageing;
- **.6** weathering;
- **.7** flex cracking;
- **.8** abrasion;
- **.9** coating adhesion;
- **.10** oil resistance;
- **.11** elongation at break;
- **.12** piercing strength;
- **.13** ozone resistance;
- **.14** gas permeability;
- **.15** seam strength; and
- **.16** ultraviolet light resistance.

Mooring out test

7.2.15 The inflated rescue boat should be subjected to the tests indicated in 5.5.

---

* Refer to the recommendations of the International Organization for Standardization, in particular publication ISO 15372 *Ships and marine technology – Inflatable rescue boats – Coated fabrics for inflatable chambers*.

Detailed inspection

7.2.16 The inflated rescue boat complete in all respects should be fully inflated in the manufacturer's works and subjected to detailed inspection to ensure that all the requirements are fulfilled.

7.3 Rigid/inflated rescue boats

7.3.1 Rigid/inflated rescue boats should be subjected to the tests prescribed in 6.2 (for hull), 7.2.14 (for inflated part), 6.4.1, 6.6.1, 6.7.2, 6.9.1 to 6.9.4, 6.10 (except 6.10.1) to 6.12, 7.1.2 to 7.1.8, 7.2.2 to 7.2.11, 7.2.15 and 7.2.16.

7.3.2 The tests prescribed in 7.2.8, 7.2.9 and 7.2.15 do not apply to rigid/inflated rescue boats if the boat has its waterline below the lower side of the inflated tube.

7.4 Rigid fast rescue boats

7.4.1 Rigid fast rescue boats should be subjected to the tests prescribed in 6.2 to 6.12 (except 6.3, 6.4.2, 6.5, 6.6.2, 6.7.1, 6.9.5, 6.9.6, 6.10.1), 6.14 (if a rigid fast rescue boat is self-righting), 7.1.2 to 7.1.4, 7.1.6, 7.1.7 (if a rigid fast rescue boat is not self-righting), 7.1.8, 7.1.9 and 7.2.4.2. In the case of open fast rescue boats, the self-righting test should only be done in the light condition, and 6.14.1.1, 6.14.3, 6.14.4, and 6.14.5 are not applicable. With regard to 6.14.2, a boat fitted with a helmsman's emergency release switch should be considered to be arranged to stop automatically when inverted.

7.4.2 *Operation tests*

Operation of engine and fuel consumption test

7.4.2.1 The boat should be loaded with weights equal to the mass of its equipment and the number of persons for which the boat is to be approved. The engine should be started and the boat manoeuvred for a period of at least 4 h to demonstrate satisfactory operation.

7.4.2.2 The boat should be run at a speed of not less than 8 knots with a full complement of persons and equipment and 20 knots with a crew of 3 persons for a period which is sufficient to ascertain the fuel consumption and to establish that the fuel tank has the required capacity.

7.5 Inflated fast rescue boats

Inflated fast rescue boats should be subjected to the tests prescribed in 6.4.1, 6.6.1, 6.7.2, 6.9.1 to 6.9.4, 6.10 (except 6.10.1), 6.11, 6.12, 6.14 (if inflated fast rescue boat is self-righting), 7.1.2, 7.1.3, 7.1.6 (if inflated fast rescue boat is equipped with an outboard motor), 7.1.7 (if inflated fast rescue boat is not self-righting), 7.1.8, 7.2.2 to 7.2.16 and 7.4.2.

7.6 Rigid/inflated fast rescue boats

Rigid/inflated fast rescue boats should be subjected to the tests prescribed in 6.2 (for hull), 7.2.14 (for inflated part), 6.4.1, 6.6.1, 6.7.2, 6.9.1 to 6.9.4, 6.10 (except 6.10.1) to 6.12, 6.14 (if rigid/inflated fast rescue boat is self-righting), 7.1.2 to 7.1.4, 7.1.6 (if rigid/inflated rescue boat is equipped with an outboard motor), 7.1.7 (if rigid/inflated fast rescue boat is not self-righting), 7.1.8, 7.2.2 to 7.2.11, 7.2.15, 7.2.16, 7.3.2 and 7.4.2.

7.7 Outboard motors for rescue boats

7.7.1 When the rescue boats are fitted with outboard motors, the following tests should be applied to the motor in place of those tests specified in 6.10.

Power test

7.7.2 The motor, fitted with a suitable propeller, should be placed in a test rig such that the propeller is completely submerged in a water tank, simulating service conditions.

7.7.3 The motor should be run at the maximum continuous rated speed using the maximum power obtainable for 20 min, and should not overheat or be damaged.

Water drench test

7.7.4 The motor protective cover should be removed and the motor thoroughly drenched with water, by hose, except for the intake to the carburettor. The motor should be started and run at speed for at least 5 min while it is still being drenched. The motor should not falter or be damaged by this test.

Hot start test

7.7.5 While still in the test rig referred to in 7.7.2, the motor should be run at idling speed in order to heat up the cylinder block. At the maximum temperature achievable, the motor should be stopped and immediately restarted. This test should be carried out at least twice. The motor should not fail to restart.

Manual start test

7.7.6 The motor should be started at ambient temperature by manual means. The means should be either a manual automatic-rewind system or a pull cord round the top flywheel of the motor. The motor should be started twice within 2 min of commencement of the start procedure.

7.7.7 The motor should be run until normal operating temperatures are reached, then it should be stopped and started manually twice within 2 min, in accordance with 7.7.6.

Cold start test

7.7.8 The motor, together with the fuel, fuel lines and battery, should be placed in a chamber at a temperature of –15°C and allowed to remain until the temperature of all parts has reached the temperature of the chamber. The temperature of the fuel, battery and motor should be measured for this test. The motor should be started twice, within 2 min of commencement of the start procedure, and allowed to run long enough to demonstrate that it runs at operating speed. It is recommended that this period should not exceed 15 s.

7.7.9 Where, in the opinion of the Administration, having regard to the particular voyages in which the ship carrying the boat is constantly engaged, a lower temperature is appropriate, that lower temperature should be substituted for –15°C in 7.7.8 for the cold start test.

Engine-out-of-water test

7.7.10 The engine should be operated for at least 5 min at idling speed under conditions simulating normal storage. The engine should not be damaged as a result of this test.

Engine inversion test (for engines destined for fast rescue boats only)

7.7.11 The engine and its fuel tank should be mounted on a frame that is arranged to rotate about an axis equivalent to the longitudinal axis of the boat at the height of the boat transom. The propeller should be in a water basin to the height of the cavitation plate. The engine should then be subjected to the test procedure specified in paragraphs 6.14.7.1 through 6.14.7.13, and then dismantled for examination. With regard to 6.14.7.9, the engine should be stopped automatically or by the helmsman's emergency release switch when inverted. During these tests, the engine should not overheat or fail to operate or leak more than 250 mℓ of oil during any one inversion. When examined after being dismantled, the engine should show no evidence of overheating or excessive wear.

8 Launching and embarkation appliances

8.1 Testing of davits and launching appliances

8.1.1 For lifeboats other than free-fall lifeboats, davits and launching appliances, except the winch brakes, should be subjected to a static proof load of 2.2 times their maximum working load. With the load at the full outboard position, the load should be swung through an arc of approximately 10° to each side of vertical in the intended fore and aft plane. The test should be done first in the upright position, followed by tests simulating a shipboard condition of list of 20° both inboard and outboard. There should be no evidence of significant deformation or other damage as a result of this test. For free-fall lifeboats, the launching appliances for lowering a free-fall lifeboat by falls, except the winch brakes, should be subjected to a static proof load of 2.2 times the maximum working load at the full outboard position. The launching ramp and its connection to the release mechanism should be subjected to a static proof load of 2.2 times the maximum working load. There should be no evidence of significant deformation or other damage as a result of this test.

8.1.2 For lifeboats other than free-fall lifeboats, a mass equal to 1.1 times the maximum working load should be suspended from the lifting points with the launching appliance in the upright position. The load should be moved from the full inboard to the full outboard position using the means of operation that is used on the ship. The test should be repeated with the launching appliance positioned to simulate a combined 20° inboard list and 10° trim. All the tests should be repeated with a mass equal to that of a fully equipped lifeboat, without persons, or the lightest survival craft intended for use with the davit to ensure the satisfactory functioning of the davit under very light load conditions. The appliance should successfully lower the load under all of the conditions, and there should be no evidence of significant deformation or other damage as a result of the tests. For freefall lifeboats, a mass equal to 1.1 times the maximum working load should be suspended from the lifting points. The load should be moved from the full inboard to the full outboard position using the means of operation that is to be used on the ship. The test should be repeated with a mass equal to that of the fully equipped lifeboat, without persons, to ensure the satisfactory functioning of the appliance under light load conditions. The appliance should successfully lower the load under both conditions and there should be no evidence of significant deformation or other damage as a result of the tests.

8.1.3 A mass equal to 1.1 times the maximum working load should be suspended from the lifting points with the launching appliance in the upright position. The load should be moved from the full inboard to the full outboard position using the means of operation that is used on the ship. The appliance should successfully move the maximum designed hoisting load from the outboard to the inboard position without causing permanent deformation or other damage.

8.1.4 Winch drums should be wound to the maximum number of turns permitted and a static test load of 1.5 times the maximum working load should be applied and held by the brake. This load should then be lowered for at least one complete revolution of the barrel shaft. A test load of 1.1 times the maximum working load should then be lowered at maximum lowering speed through a distance of at least 3 m and stopped by applying the hand brake sharply. For a lifeboat or rescue boat launching appliance, the test load should drop not more than 1 m when the brake is applied. For a fast rescue boat launching appliance, the test load should come to a rapid, but gradual stop, and the dynamical force induced in the wire should not exceed 0.5 times the working load of the launching appliance. This test should be repeated a number of times. If the winch design incorporates an exposed brake, one of these tests should be carried out with the brake wetted, but in this case the stopping distance may be exceeded. The various tests should achieve a cumulative lowering distance of at least 150 m. Operation of the winch with a load of a mass equal to that of a fully equipped lifeboat, without persons, or the lightest survival craft intended for use with the winch, should also be demonstrated.

8.1.5 It should be demonstrated that a winch intended for use with a rescue boat is capable of recovering the rescue boat with the number of persons for which it is to be approved and its equipment or an equivalent mass at a rate of not less than 0.3 m/s or 0.8 m/s in the case of a launching appliance for a fast rescue boat.

8.1.6 The hand operation of the winch should be demonstrated. If the winch is designed for quick recovery by hand with no load, this should be demonstrated with a load of 1.5 times the mass of the empty lifting arrangements.*

8.1.7 Following completion of the tests the winch should be stripped for inspection. These tests and the inspection should normally be witnessed by a representative of the Administration.

* This paragraph does not apply to free-fall lifeboats.

8.1.8 The fast rescue boat launching appliance should be demonstrated in a sea state associated with a force 6 wind on the Beaufort scale, in association with a significant wave height of at least 3 m. The test should include launching and recovery of a fast rescue boat and demonstrate:

- **.1** satisfactory operation of the device to dampen forces and oscillations due to interaction with the waves;
- **.2** satisfactory operation of the winch brake; and
- **.3** satisfactory operation of the tensioning device.

8.2 Davit-launched liferaft automatic release hook test

Definitions

8.2.1 In this section and in 6.2.1 to 6.2.7 of Part 2, the following definitions apply:

- **.1** *Actuating force* means the force required to set the actuating mechanism.
- **.2** *Actuating mechanism* means the mechanism which, when operated, allows the liferaft to be released automatically.
- **.3** *Automatic release mechanism* means the mechanism which opens the hook automatically to release the liferaft.
- **.4** *Hook* means a hook to be used for the launching of liferafts which can be activated to automatically release the liferaft when it is waterborne.
- **.5** *Load limit for automatic release* means the minimum load at which the automatic release mechanism opens the hook and automatically and completely releases the liferaft.
- **.6** *Manual release force* means the force required at the actuating mechanism to release the hook manually.
- **.7** *Safe working load* means the load for which the hook is to be approved.
- **.8** *Securing force* means the force required to close the hook manually.
- **.9** *Testing establishment* means an establishment accepted by the Administration having the equipment and the qualifications necessary for the testing and approval of liferaft release hooks.

Hook and documents required for tests

8.2.2 The following should be submitted to the testing establishment for the prototype testing of the hook:

.1 two hooks which have been passed for delivery; and

.2 a functional description of the hook together with any other documents necessary to carry out the tests.

Corrosion resistance test

8.2.3 Two hooks should be submitted to a corrosion resistance test which should be made in a salt mist chamber in accordance with ISO 3768:1976 for 1,000 h or equivalent national standard. Any corrosion effects and other damage to the hooks should be recorded.

8.2.4 Both hooks should then be subjected five times to the tests required by 8.2.5 to 8.2.17.

Load test

8.2.5 The maximum load on the hook to allow for automatic release loads should be determined as follows:

.1 the hook should be loaded with a mass of 200 kg and the actuating mechanism set;

.2 the load should be reduced gradually in stages until the hook releases automatically, but at not more than 30 kg, to establish load F; and

.3 the load F should be measured and recorded. The minimum allowable load F is the minimum obtained at release which should be not less than 5 kg.

8.2.6 The hook should be loaded with a mass of 200 kg and the actuating mechanism set. The hook should then be subjected to cyclic loading between 30 kg and 200 kg using a frequency of 1 ± 0.2 Hz. The hook should not release before 300 cycles. The number of cycles at which the hook opened or whether the test was discontinued at 300 cycles should be recorded.

8.2.7 The hook should then be reloaded to 200 kg and the actuating mechanism set. The hook should be subjected to a cyclic loading, the upper limit of which is +200 kg and the lower limit being F_1, using a frequency of 1 ± 0.2 Hz. The automatic release mechanism should operate within three cycles. The number of cycles at which the hook opened or whether the test was discontinued after three cycles should be recorded. F_1 is to be taken as the minimum load on the hook to allow for automatic release, as established in 8.2.5.2, reduced by 2 kg.

8.2.8 The hook should be attached to a short wire rope fall, approximately 1.5 m, and loaded with a mass of 10 kg. The weight should be secured and then lifted 1 m. From this position it should be released to perform a free fall before it is abruptly stopped by the wire rope fall. The hook should not release as a result of this test.

8.2.9 The automatic release hook should be attached to a test load equal to 1.1 times the safe working load, with the actuating mechanism in the locked position. The load should be raised to a height of at least 6 m and then be lowered at a speed of 0.6 m/s. When the load is 1.5 m above the ground or water surface, the actuating mechanism should be set for automatic release, and the lowering completed. The automatic release hook should release the load when it strikes the ground or water surface. The test should be repeated with a test load equal to 2.2 times the safe working load.

8.2.10 The automatic release hook should be attached to a test load of 1.1 times its maximum working load using an approved launching appliance. The test load should be lowered at maximum lowering speed through a distance of at least 3 m and stopped by applying the hand brake sharply. This test should be conducted twice, once with the release mechanism set for automatic release, and again with the mechanism set to closed. The release mechanism should not open in either test.

8.2.11 The hook should be loaded to 0%, 25%, 50%, 75% and 100% of the safe working load of the hook. At each load level, the actuating force required at the actuating mechanism should be measured and recorded. The actuating force should be between 150 N and 250 N if the hook is lanyard-operated, or the action required to set the actuating mechanism should be readily performed by a single person without difficulty.

8.2.12 The securing force, which should be less than 120 N, should be determined with an unloaded hook and should be recorded.

8.2.13 The manual release force should be determined as follows:

.1 the hook should be loaded with a mass of 150 kg;

.2 the actuating mechanism should be set;

.3 the force required to release the hook manually should he established and recorded; and

.4 the manual release force for a mass of 150 kg on the hook should be at least 600 N for lanyard-operated designs. Alternative designs should be demonstrated to the satisfaction of the Administration to provide adequate protection from inadvertent release under load.

8.2.14 The automatic release hook should be attached to a test load equal to the mass of the lightest liferaft for which the automatic release hook is to be approved, with the actuating mechanism in the locked position (i.e., not set for automatic release). The load should then be raised so that it is clear of the ground. The actuating mechanism should be set to automatic release. This should be easily accomplished by a single person and should not release the load.

8.2.15 The hook should be released 100 times without failure by each of its modes of release using the maximum load permitting release for that mode. It should then be disassembled and the parts examined. There should be no evidence of excessive wear on any part.

8.2.16 The hook should be arranged in a cold store at –30°C to simulate operational readiness and loaded with 25 kg. A 3.5 cm thick uniform layer of icing should be built onto it by spraying cold fresh water from angles above 45° from horizontal, with intermittent pauses to let icing form. The hook should then be actuated and as a result release the hold without failure.

8.2.17 It should be demonstrated that the hook is not damaged as a result of 10 impacts at a horizontal speed of 3.5 m/s on to a structure resembling a vertical ship's side. As far as practical all sides of the hook, especially areas with exposed controls, should impact the structure. The hook should not sustain any damage which will interfere with the normal function of the hook.

Compatibility of liferaft and release hook

8.2.18 Where automatic release hooks are supplied for use with liferafts made by different manufacturers, operational tests with each type and size of lifting or attachment fitting used by the different manufacturers of the liferafts should be carried out before the particular combination of liferaft and release hook is accepted by the Administration.

9 Line-throwing appliances

9.1 Test for pyrotechnics

Rockets used in line-throwing appliances should be subjected to the tests prescribed in 4.3.1, 4.3.3, 4.4, 4.5.1 (if appropriate), 4.5.5 and 4.5.6.

9.2 Function test

Three projectiles should be fired connected to a line and should carry the line at least 230 m in calm conditions. The lateral deflection from the line of firing should not exceed 10% of the length of flight of the projectile. If the projectile is fired using an explosive charge, then one of the projectiles should be fired using double the normal charge.

9.3 Line tensile test

The line should be subjected to a tensile test and should have a breaking strain of not less than 2 kN.

9.4 Visual examination

It should be established by visual examination that the appliance:

.1 is marked with clear and precise instructions on how it should be operated; and

.2 is marked with a means of determining its age.

9.5 Temperature test

Three individual units, consisting of projectile, firing system and line should be subjected to the temperature cycling prescribed in 4.2.1 and one specimen subjected to each of the tests prescribed in 4.2.2, 4.2.3 and 4.2.4.

10 Position-indicating lights for life-saving appliances

10.1 Survival craft and rescue boats light tests

10.1.1 Twelve liferaft canopy lights, lifeboat enclosure or lifeboat cover lights, as the case may be, and twelve survival craft interior lights should be subjected to the temperature cycling as prescribed in 1.2.1. If the same type of light is used for both canopy, enclosure or cover and interior, only twelve lights of that type need to be tested. If the lifeboat enclosure light, the lifeboat cover light or the lifeboat internal light is connected to the lifeboat's electrical network and can be supplied with electrical power from any one of the lifeboat's batteries as well as from the lifeboat's engine-driven generator set, the light should only be subject to the test as far as practicable.

10.1.2 In the case of seawater cell power sources, four survival craft lights of each type should, following at least ten complete temperature cycles, be taken from a stowage temperature of –30°C and be operated immersed in seawater at a temperature of –1°C; four of each type should be taken from a stowage temperature of +65°C and be operated immersed in seawater at a temperature of +30°C; and four of each type should be taken from ordinary room conditions and operated immersed in fresh water at ambient temperature. The canopy, enclosure or cover lights should be of white colour and should provide a luminous intensity of not less than 4.3 cd in all directions of the upper hemisphere for a period of not less than 12 h (see 10.4). The interior lights should provide sufficient luminous intensity to read survival instructions and equipment instructions for a period of not less than 12 h.

10.1.3 In the case of dry cell power sources, provided they will not come into contact with seawater, four survival craft lights of each type should, following at least ten complete temperature cycles be operated at an air temperature of –30°C, four of each type at an air temperature of +65°C, and four of each type at ambient temperature. The canopy, enclosure or cover lights should be white in colour and should provide luminous intensity of not less than 4.3 cd in all directions of the upper hemisphere for a period of not less than 12 h (see 10.4). The interior lights should provide an arithmetic mean luminous intensity of not less than 0.5 cd when measured over the entire upper hemisphere to permit reading of survival instructions and equipment instructions for a period of not less than 12 h.

10.1.4 In the case of a flashing light, it should be established that the rate of flashing for the 12 h operative period is not less than 50 flashes and not more than 70 flashes per minute and the effective luminous intensity is at least 4.3 cd (see 10.4).

10.2 Lifebuoy self-igniting light tests

10.2.1 Three self-igniting lights should be subjected to temperature cycling as prescribed in 1.2.1.

10.2.2 After at least ten complete temperature cycles, one self-igniting light should then be taken from a stowage temperature of –30°C and operated immersed in seawater at a temperature of –1°C, and another should be taken from a stowage temperature of +65°C and operated immersed in seawater at a temperature of +30°C. Both lights should be of white colour and should continue to provide a luminous intensity of not less than 2 cd in all directions of the upper hemisphere or, in the case of a flashing light, flash at a rate of not less than 50 flashes and not more than 70 flashes per minute with at least the corresponding effective luminous intensity for a period of not less than 2 h (see 10.4).

At the end of the first hour of operation the lights should be immersed to a depth of 1 m for 1 min. The lights should not be extinguished and should continue operating for at least an hour longer.

10.2.3 A self-igniting light should be subjected to two drop tests into water as prescribed in 1.3. The light should be dropped twice, first by itself and then attached to a lifebuoy. The light should operate satisfactorily after each drop.

10.2.4 A self-igniting light should be allowed to float in water in its normal operating position for 24 h. If the light is an electric light, it should be disassembled at the end of the test and examined for the presence of water. There should be no evidence of water inside the light.

10.2.5 The remaining self-igniting light, which has been subjected to the test in 10.2.1 should be immersed horizontally under 300 mm of water for 24 h. If the light is an electric light, it should be dismantled at the end of the test and examined for the presence of water. There should be no evidence of water inside the light.

10.2.6 If a self-igniting light has a lens, the light should be cooled to –18°C and dropped twice from a height of 1 m onto a rigidly mounted steel plate or concrete surface. The distance should be measured from the top of the lens to the impact surface. The light should strike the surface on the top centre of the lens. The lens should not break or crack.

10.2.7 A self-igniting light should be placed on its side on a rigid surface and a steel sphere having a mass of 500 g should be dropped from a height of 1.3 m onto the case three times. The sphere should strike the case near its centre on one drop, approximately 12 mm from one end of the case on another drop and approximately 12 mm from the other end of the case on the third drop. The case should not break or crack, or be distorted in a way that would affect its watertightness.

10.2.8 A force of 225 N should be applied to the fitting that attaches the light to a lifebuoy. Neither the fitting nor the light should be damaged as a result of this test.

10.3 Lifejacket light tests

10.3.1 Twelve lifejacket lights should be subjected to temperature cycling as prescribed in 1.2.1.

10.3.2 After at least ten temperature cycles, four of these lifejacket lights should be taken from a stowage temperature of −30°C and then be operated immersed in seawater at a temperature of −1°C. Four should be taken from a stowage temperature of +65°C and then immersed in seawater at a temperature of +30°C and four should be taken from ordinary room conditions and operated immersed in freshwater at ambient temperature. Water-activated lights should commence functioning within 2 min and have reached a luminous intensity of 0.75 cd within 5 min in seawater. In fresh water a luminous intensity of 0.75 cd should have been attained within 10 min. At least 11 out of the 12 lights, which should all be of white colour, should continue to provide a luminous intensity of not less than 0.75 cd in all directions of the upper hemisphere for a period of at least 8 h.

10.3.3 One light attached to a lifejacket should be subjected to a drop test from 4.5 m as prescribed in 2.8.8. The light should not suffer damage, should not be dislodged from the lifejacket and should be switched on and seen to be illuminated and conspicuous whilst the test subject is still in the water.

10.3.4 One light should be dropped from a height of 2 m onto a rigidly mounted steel plate or concrete surface. The light should not suffer damage and should be capable of providing a luminous intensity of not less than 0.75 cd for a period of at least eight hours when operated immersed in freshwater at ambient temperature.

10.3.5 In the case of a flashing light it should be established that:

- **.1** the light can be operated by a manual switch;
- **.2** the rate of flashing is not less than 50 flashes and not more than 70 flashes per minute; and
- **.3** the effective luminous intensity is at least 0.75 cd (see 10.4).

10.4 Common tests for all position-indicating lights

(additional lights are required to carry out the environmental tests)

10.4.1 *Vibration test*

Regulation: IEC 60945: 2002, paragraph 8.7

Test procedure

One unit should be subjected to a vibration test according to IEC 60945: 2002, paragraph 8.7.

Acceptance criteria

The lights should function after the test.

10.4.2 *Mould growth test*

Regulation: LSA Code, paragraph 1.2.2.4

Test procedure

One unit should be subjected to the mould growth test.

> *Note:* The mould growth test may be waived where the manufacturer is able to produce evidence that the external materials employed will satisfy the test.

The light should be inoculated by spraying with an aqueous suspension of mould spores containing all the following cultures:

Aspergillus niger;
Aspergillus terreus;
Aureobasidium pullulans;
Paecilomyces variotii;
Penicillium funiculosum;
Penicillium ochro-chloron;
Scopulariopsis brevicaulis; and
Trichoderma viride.

The light should then be placed in a mould growth chamber which shall be maintained at a temperature of 29 ± 1°C and a relative humidity of not less than 95%. The period of incubation should be 28 days. After this period the light should be inspected.

Acceptance criteria

The light should be rot-proof and not be unduly affected by fungal attack. There shall be no mould growth visible to the naked eye and the light should function after the test.

10.4.3 *Switch arrangement test*

Test procedure

One unit should be subjected to the switch arrangement test. A person, wearing immersion suit gloves, must be able to switch the light in its normal operational position on and off three times.

Acceptance criteria

The light must function properly.

10.4.4 *Corrosion and seawater resistance test*

Test procedure

One unit should be subjected to a corrosion and seawater resistance test according to IEC 60945: 2002, paragraph 8.12.

Notes: .1 If there are no exposed metal parts the corrosion and seawater resistance test need not be conducted.

.2 The corrosion and seawater resistance test may be waived where the manufacturer is able to produce evidence that the external metals employed will satisfy the test.

Acceptance criteria

There should be no undue deterioration of metal parts and the unit should function.

10.4.5 *Solar radiation test*

(not for survival craft interior and lifejacket lights)

Test procedure

One unit should be subjected to a solar radiation test according to IEC 60945: 2002, paragraph 8.10.

Note: The solar radiation test may be waived where the manufacturer is able to produce evidence that the materials employed will satisfy the test, i.e., UV stabilized.

Acceptance criteria

The mechanical properties and labels of the unit should be resistant to harmful deterioration by sunlight. The unit should function after the test.

10.4.6 *Test for oil resistance*

(not for survival craft interior lights)

Test procedure

One unit should be subjected to the oil resistance test according to IEC 60945: 2002, paragraph 8.11.

Acceptance criteria

After this test the unit should not be unduly affected by oil and should show no sign of damage such as shrinking, cracking, swelling, dissolution or change of mechanical qualities. The light should function after the test.

10.4.7 *Rain test, and watertightness test*

Test procedure

One unit should be subjected to a rain test according to IEC 60945: 2002, paragraph 8.8. After having passed the rain test, the unit and the complete power source should be immersed horizontally under not less than 300 mm of fresh water for at least 24 h.

Acceptance criteria

The unit should comply with the requirements of IEC 60945: 2002, paragraph 8.8.2, and should function after the rain test. Additionally, after the watertightness test the unit should function and there should be no evidence of water inside the unit.

10.4.8 *Fire test*

(not for survival craft interior lights)

Test procedure

One unit should be subjected to a fire test. A test pan at least 30 cm × 35 cm × 6 cm should be placed in an essentially draught-free area. Water should be put in the bottom of the test pan to a depth of not less than 1 cm followed by enough petrol to make a minimum total depth of not less than 4 cm. The petrol should then be ignited and allowed to burn freely for at least 30 s. The unit should then be moved through the flames, facing them, with the unit's light not more than 25 cm above the top edge of the test pan so that the duration of exposure to the flames is at least 2 s.

Acceptance criteria

The unit should not sustain burning or continue melting after being totally enveloped in a fire for a period of at least 2 s and after being removed from the flames. The unit should function after the test.

10.4.9 *Measurement of luminous intensity*

Test procedure

If the voltage at five minutes of operation is lower than the recorded voltage at the end of life, it is permissable to use a lamp from the same build standard for the light output test. Using the lowest recorded voltage a light output test can be carried out as described below. The voltage of the specified number of test units should be monitored continuously for the specified time. To make sure that all the test units provide a luminous intensity of not less than the specified luminous intensity in all directions of the upper hemisphere after the specified time of operation, the following test should be performed.

It must be demonstrated that at least one light from each of the specified temperature ranges reaches the required luminous intensity in all directions of the upper hemisphere when using a photometer which is calibrated to the photometric standards of the appropriate national or State standards institute. (Note: CIE Publication No. 70 contains further information.) The lowest voltage light of the cold temperature test sample lot, the highest voltage light of the high temperature test sample lot and the mean voltage light of the ambient temperature sample lot should be selected. These three lights must be used for the light output tests. In the event that a lamp filament burns out during the light output test, a second light from the same performance test lot may be used.

Luminous intensity should be measured by a photometer directed at the centre of the light source with the test light on a rotating table. Luminous intensity should be measured in a horizontal direction at the level of the centre of the light source and continuously recorded through a 360° rotation. These measurements should be taken in the azimuth angles at 5° intervals above the horizon up to the single measurement at 90° (vertical). Luminous intensity should then be measured in a vertical direction, beginning at the centre of the light source at the point of lowest recorded light output, and continuously recorded through an arc of 180°.

Acceptance criteria

The test lights should continue to provide a luminous intensity of not less than the specified intensity in all directions of the upper hemisphere for a period of at least the specified time. All measured data of luminous intensity and voltage should be documented. In the case of a flashing light, it should be established that the rate of flashing for the specified operating period is not less than 50 flashes and not more than 70 flashes per minute and that the effective luminous intensity is at least the minimum specified intensity in all directions of the upper hemisphere. The effective luminous intensity is to be found from the formula:

$$\left(\frac{\int_{t_1}^{t_2} I dt}{0.2 + (t_2 - t_1)} \right)_{max}$$

where:

I is the instantaneous intensity, 0.2 is the Blondel–Rey constant and t_1 and t_2 are time-limits of integration in seconds.

Flashing lights with a flash duration of not less than 0.3 s may be considered as fixed/steady lights for the measurement of their luminous intensity. Such lights should provide the required luminous intensity in all directions of the upper hemisphere. The time interval between switching on and reaching the required luminous intensity (incandescence time) and all time spent below the required luminous intensity when the light switches off should be disregarded (see figure 10.4.1.).

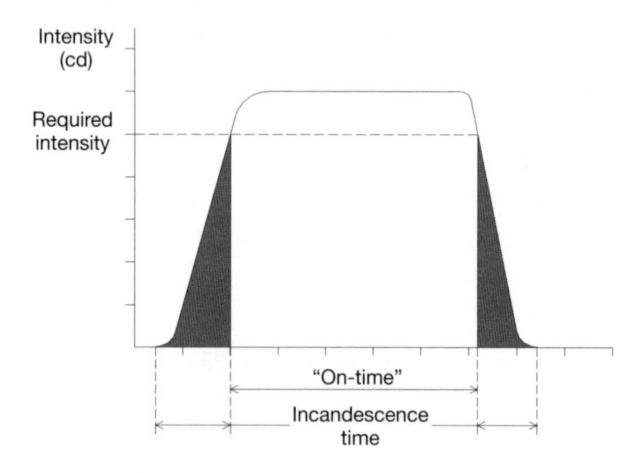

Figure 10.4.1 – *"On-time" measurement diagram*

10.4.10 *Chromaticity*

Test procedure

One unit should be tested for chromaticity to determine that it lies within the boundaries of the area "white" of the diagram specified for each colour by the International Commission on Illumination (CIE). The chromaticity of the light should be measured by means of colorimetric measurement equipment which is calibrated to the appropriate national or State Standards Institute. (Note: CIE Publication No. 15.2 contains further information.) Measurements on at least four points of the upper hemisphere should be taken.

Acceptance criteria

The measured chromaticity co-ordinates should fall within the boundaries of the area of the diagram, as per CIE. The boundaries of the area for white lights are given by the following corner co-ordinates:

x	0.500	0.500	0.440	0.300	0.300	0.440
y	0.382	0.440	0.433	0.344	0.278	0.382

(International Standard on Colours of Light Signals, with colour tables to be developed by CIE.)

11 Hydrostatic release units

11.1 Visual and dimensional examination

Two samples of hydrostatic release units should be given a visual and dimensional examination. If the devices conform with the manufacturer's drawings and specifications, they should be accepted and assembled for further testing under the technical and performance tests as prescribed in 11.2 and 11.3.

11.2 Technical tests

Each hydrostatic release unit should undergo all the following technical tests. No parts should be renewed or repaired between the tests. The tests should be conducted in the following sequence:

.1 *Corrosion resistance test*

A hydrostatic release unit should be exposed to a salt water spray test (5% natrium chloride solution) at a temperature of 35 ± 3°C for 160 h without interruption. After completion of the test the hydrostatic release unit should show no corrosion which could affect its efficient functioning and should then be subjected to the following tests after which it should continue to function efficiently.

.2 *Temperature test*

The hydrostatic release units should then be subjected to the temperature cycling prescribed in 1.2.1. Following temperature cycling as prescribed in 1.2.1, one hydrostatic release unit should be taken from a stowage temperature of −30°C and should then operate in seawater at a temperature of −1°C. The other hydrostatic release unit should be taken from a stowage temperature of +65°C, and should then operate in seawater at a temperature of +30°C.

.3 *Submergence and manual release tests*

The hydrostatic release unit should then be tested by applying a buoyant load equal to its designed capacity while the device is submerged in water or in a water-filled pressure testing tank. It should release at a depth of not more than 4 m. On completion of these tests and resetting, the hydrostatic release unit should be capable of being released manually if it is designed to allow manual release of the unit. It should then be opened for inspection and should show no significant signs of corrosion or degradation.

.4 *Strength test*

After reassembly the hydrostatic release unit, if forming part of the painter system, should be subjected to a tensile test of at least 10 kN for a period of 30 min. If the release unit is to be fitted to a liferaft for more than 25 persons it should be subjected to a tensile test of at least 15 kN. After the tensile test the unit, if designed to allow manual release, should then be capable of being operated manually.

.5 *Technical tests on the membrane*

The following tests should be carried out on the membrane:

.5.1 *Test of resistance to cold*

Number of specimens:	2 membranes
Temperature:	−30°C
Exposure time:	30 min
Flex testing:	180° with both inside and outside stretched
Requirements:	The membranes should show no visible cracking.

.5.2 *Test of resistance to heat*

Number of specimens:	2 membranes
Temperature:	+65°C
Exposure time:	7 days
Requirements:	The membranes should show no visible cracking.

.5.3 *Test for surface resistance to oil*

Number of specimens:	2 membranes
Temperature:	+18°C to +20°C
Type of oil:	A mineral oil meeting the following requirements: Aniline point: 120 ± 5°C Flashpoint: minimum 240°C Viscosity: 10–25 cSt at 99.0°C
The following oils may be used:	ASTM Oil No. 1 ASTM Oil No. 5 ISO Oil No. 1
Testing period:	3 h on each side
Requirements:	The material should show no deterioration.

.5.4 *Resistance to seawater*
Two membranes should be immersed for 7 days in 5% natrium chloride solution:

Test temperature:	+18°C to +20°C
Requirements:	The material should show no deterioration.

.5.5 *Resistance to detergents*
Two membranes should be immersed for 7 days in detergents commonly used on board ship:

Test temperature:	+18°C to +20°C
Requirements:	The membranes should show no signs of deterioration.

.6 Solar radiation test

One unit should be subjected to a solar radiation test to paragraph 8.10 of IEC 60945: 2002.

Note: The solar radiation test may be waived where the manufacturer is able to produce evidence that the materials employed will satisfy the test, i.e., UV stabilized.

11.3 Performance test

11.3.1 This test should be performed using the smallest and the largest liferafts with which the hydrostatic release unit may be used. If the occupant range between the smallest and largest liferaft exceeds 25 persons, then the intermediate size liferaft should also be tested. The liferaft should be placed horizontally on a rack or platform of sufficient weight to submerge the liferaft. The hydrostatic release unit and painter should be installed as aboard ship.

11.3.2 The following tests should be carried out in a suitable depth of water. The platform on which the liferaft is mounted should be lowered into the water as follows:

.1 horizontal;

.2 tilted 45° and then 100° with the hydrostatic release unit at the upper side;

.3 tilted 45° and then 100° with the hydrostatic release unit at the lower side; and

.4 vertically.

Under these conditions the hydrostatic release unit should release the liferaft at a depth of less than 4 m.

12 Marine evacuation systems

12.1 Materials

Materials used in the construction of marine evacuation systems are to be tested to the standards laid down in 5.17.13, where applicable.

12.2 Marine evacuation system container

12.2.1 It should be demonstrated that the passage and platform if fitted, or liferafts in any other case, can be deployed from the container by one person in a sequence prescribed in the manufacturer's instruction. If more than one action is necessary to operate the system means should be provided to prevent incorrect operation.

12.2.2 A static load of 2.2 times the maximum load on the system should be applied to its structural attachment to the ship for a period of 30 min. This static load is to be equivalent to the calculated load imposed by the maximum number and size of fully loaded liferafts for which the system is designed, attached to the loaded platform with the ship moving through the water at 3 knots against a head wind of force 10 on the Beaufort scale. There should be no evidence of significant deformation or other damage as a result of this factory test.

12.2.3 The exterior of the container as installed should be hose tested in a similar manner to the canopy closure test in 5.12 to ensure that it is reasonably weathertight to prevent the ingress of water. Alternatively, when hose testing is required to verify the tightness of the structures the minimum pressure in the hose, at least equal to 2 bar, is to be applied at a maximum distance of 1.5 m. The nozzle diameter is not to be less than 12 mm.

12.2.4 The release and securing arrangements for any internal or external doors are to be satisfactorily tested by five dry release operations carried out consecutively.

12.2.5 It should be demonstrated by two dry deployments of the system, with the container angled back to simulate an unfavourable trim of up to 10° and list of up to 20° either way, that the outer door, the passage and platform (if fitted) will not suffer damage which render it unusable for its intended purpose.

12.3 Marine evacuation passage

12.3.1 For an inclined inflated passage the following requirements are to be complied with:

- **.1** a fully inflated passage should be arranged on a solid base at the height at which it is to be stowed on board. When loaded at mid-length with a weight of 150 kg for each single slide path the passage must not become unduly distorted;

.2 a fully inflated passage should be subjected to individual sliding operations twice the number for which it is to be certificated. For this test actual persons of varied physique and weight should be used. On completion the slide path must remain in a serviceable condition;

.3 it should be demonstrated using actual persons that the loss of pressure in any one section of the passage will not limit its use as a means of evacuation;

.4 a static load of 2.2 times the maximum to which the system is to be designed, in accordance with 12.2.2, should be applied for a period of 30 min to the connection between the passage and the container. On completion there must be no signs of any fracture or stranding of its connections;

.5 the uninflated passage with its gas cylinders should be placed in a cold chamber at a temperature of −30°C. After a period of not less than 24 h at this temperature the chute should reach its working pressure within 5 min. The components must show no sign of cracking, seam slippage or other defects;

.6 the uninflated passage with its gas cylinders should be placed in a hot chamber at a temperature of +65°C for not less than 7 h. On inflation the pressure relief valves on the passage should be of sufficient capacity to prevent pressure in excess of twice the designed working pressure;

.7 it should be demonstrated with at least 10 sliding operations on a slide path thoroughly wetted with water to simulate wet weather conditions, that the speed of descent is not excessive or dangerous; and

.8 a pressure test is to be carried out in accordance with 5.17.7 and 5.17.8.

12.3.2 For vertical passage systems the following requirements are to be complied with:

.1 the passage(s) should be subjected to individual descent operations twice the number for which it is to be certificated. For this test actual persons of varied physique and weight should be used. On completion the passage path should remain in a serviceable condition;

.2 a load of 2.2 times the maximum to which the system is to be designed, in accordance with 12.2.2, should be applied for a period of 30 min to the connection between the passage and the container. On completion there must be no signs of any fracture or stranding of its connections;

.3 the stowed passage should be placed in a cold chamber at a temperature of –30°C. After a period of 24 h at this temperature the passage should show no signs of cracking or other defects; and

.4 it should be demonstrated with at least 10 descent operations, in the case of open vertical passages with the path thoroughly wetted with water to simulate wet weather conditions, that the speed of descent is not excessive or dangerous.

12.4 Marine evacuation platform, if fitted

12.4.1 The platform should be inflated and loaded with the number of persons carried in accordance with the number specified by paragraph 6.2.1.3.3 of the LSA Code, all wearing an approved lifejacket. Freeboards are to be measured all round, and should not be less than 300 mm.

12.4.2 It should be demonstrated that in the event of the loss of 50% of the buoyancy in the tubes the platform should be capable of supporting, with a positive freeboard all round, the number of persons specified in paragraph 6.2.1.3.3 of the LSA Code.

12.4.3 It should be demonstrated that the platform is self draining with no possibility of a build up of water.

12.4.4 The platform with its inflation system should be placed in a cold chamber at a temperature of –30°C. After a period of not less than 24 h at this temperature the platform on being inflated should achieve its normal working pressure in not more than 5 min. There should be no seam slippage, cracking or other defects on the platform, and it should be ready for use on completion of the test.

12.4.5 The platform with its inflation system should be placed in a hot chamber at a temperature of +65°C for not less than 7 h. On being inflated the pressure relief valves should be of sufficient capacity to prevent pressure in excess of twice the designed working pressure.

12.4.6 A pressure test is to be carried out in accordance with 5.17.7 and 5.17.8.

12.5 Associated inflatable liferafts

12.5.1 Liferafts used in conjunction with the marine evacuation system should conform and be prototype tested to the requirements of section 5.

12.5.2 It should be demonstrated that the liferafts can be deployed from their stowage position, and moored alongside the platform, if fitted, before being inflated, and bowsed in ready for boarding.

12.5.3 It should be demonstrated that the liferafts can be deployed from their stowage positions independently of the marine evacuation system.

12.5.4 It should be demonstrated that the liferafts will float free from their stowage positions, inflate and then break free in the event of the ship sinking.

12.5.5 If the passage is to give direct access to the liferaft(s), it should be demonstrated that it can be easily and quickly detached.

12.6 Performance

12.6.1 It should be demonstrated in harbour by full deployment of a system, including the launching and inflation of all the associated liferafts, that the system will provide a satisfactory means of evacuation. For this trial the number of persons to be used should be that for which the system is to be certificated. The various stages of this trial should be timed so as to permit the calculation of the number of persons that can be evacuated in any specified period.

12.6.2 It should be demonstrated at sea by full deployment of a system, including the launching and inflation of the associated liferafts, that the system will provide a satisfactory means of evacuation in a sea state associated with a wind of force 6 on the Beaufort scale, and in association with a significant wave height of at least 3 m. During the sea trial, a spectrum analysis of the recorded wave height shall be performed. The signal shall be high-pass filtered at 0.08 Hz to exclude any contributions from swell. The significant wave height shall be calculated based on filtered spectrum and shall not be less than 3 m. The demonstration should be carried out in accordance with the following procedures:

- **.1** Phase 1 – Initial deployment of system
  - **.1** with the ship in a simulated "dead ship" condition and the bow into the wind, the system (passage and platform or any other configuration) should be deployed in its normal design manner, and

- **.2** the platform and passage are to be observed from the ship to verify in this condition that it forms a stable evacuation system for the platform crew to descend and carry out their initial duties in preparation for evacuation;

.2 Phase 2 – Lee side trial

- **.1** the ship to be manoeuvred to place the system on the lee side and then allowed to freely drift;
- **.2** where the system employs a platform, the nominated number of the platform crew are to descend via the passage and retrieve at least two liferafts which have been launched separately;
- **.3** where the system employs a passage giving direct access to the liferaft, the nominated number of liferaft boarding crew are to descend via the passage. If additional liferafts are employed with the system, then they should be launched separately and be retrieved by the liferaft crew; and
- **.4** after the liferafts have been satisfactorily deployed, dependent upon safety consideration 20 persons in suitable protective clothing are to evacuate to the liferafts through the passage;

.3 Phase 3 – Loaded trial lee side

- **.1** the platform, if fitted, and the required number of liferafts are to be loaded to their certified capacity with weights representing 75 kg/person; and
- **.2** when loaded with the required weights the system is to be observed for a period of 30 min, with the ship free to drift, to confirm the system continues to provide a safe and stable evacuation system;

.4 Phase 4 – Loaded trial weather side

- **.1** the trials in 12.6.2.2 and 12.6.2.3 should be repeated with the system deployed on the weather side of the ship. The lee side trial and the weather side trial may be conducted in any convenient order;
- **.2** where ship manoeuvres are required to place the system on any one side, any damage or failure sustained during this manoeuvre should not constitute a failure of the system; and

.3 the system should be tested, as far as practicable, on a vessel having similar characteristics to the types of ships the equipment is to be fitted to.

13 Searchlights for lifeboats and rescue boats

13.1 Visual examination

Searchlights should be marked clearly and durably according to the requirements contained in the LSA Code, paragraphs 1.2.2.9 and 1.2.3 and additionally with the manufacturer's label.

Furthermore, on the illuminant and on the manufacturer's label, the voltage and power consumption should be marked clearly and durably.

According to the LSA Code, paragraph 1.2.2.10, searchlights should, where applicable, be provided with an electrical short circuit protection to prevent damage or injury.

With respect to the LSA Code, paragraph 4.4.6.11, means should be provided for recharging searchlight batteries.

The illuminant should be safely fitted in the searchlights; use of screwed sockets should be avoided.

Searchlights should be designed in such a way that the illuminant can be easily replaced also in darkness.

All parts of searchlights should be made of non-magnetic material.

Searchlights should be so constructed that the accumulation of condensed water in hazardous quantities is avoided.

With respect to safety precautions, searchlights should meet the relevant requirements of resolution A.694(17) and IEC 945.

13.2 Durability and resistance to environmental conditions

Temperature tests

13.2.1 Searchlights which have passed the visual examination, should be subjected to temperature tests to ensure their compliance with paragraphs 1.2.2.1 and 1.2.2.2 of the LSA Code. First they should be subjected to the dry heat test according to IEC 945 paragraph 8.2, followed by the damp heat test (8.3), the low temperature test (8.4), and thermal shock test (8.5). After these tests, searchlights should show no sign of loss of rigidity, and no sign of damage such as shrinking, cracking, swelling, dissolution or change of mechanical qualities and should be capable of being operated.

Vibration test

13.2.2 Searchlights which have passed the temperature tests, should be subjected to a vibration test according to IEC 945 paragraph 8.7 to ensure their compliance with the requirements of paragraphs 1.2.2.1 and 1.2.2.8 of the LSA Code. After the vibration test, searchlights should show no sign of damage and should be capable of being operated.

Corrosion and rain test

13.2.3 Searchlights which have passed the vibration test, should be subjected first to a corrosion test according to IEC 945 paragraph 8.12, where applicable, and second to a rain test according to IEC 945 paragraph 8.8 to ensure their compliance with the requirements of paragraphs 1.2.2.1 and 1.2.2.4 of the LSA Code. After these tests, searchlights should show no sign of damage and should be capable of being operated.

Interference

13.2.4 With respect to electrical and electromagnetic interference, searchlights should meet applicable requirements of resolution A.694(17) and of IEC 945, paragraph 9.

Power supply

13.2.5 Searchlights should be operated with 12 V or 24 V. The power supply of searchlights should meet the applicable requirements of resolution A.694(17) and of IEC 945.

13.3 Operational controls

The operational controls of searchlights should meet the requirements of resolution A.694(17) and the applicable requirements of IEC 447 and IEC 945.

Additionally, the outer parts of searchlights should not reach temperatures during operation which restrict their manual use.

13.4 Light tests

Searchlights which have passed the corrosion and rain test, and which are additionally in compliance with the requirements of 13.2.4, 13.2.5 and 13.3 above, should be subjected to the following light tests to ensure their compliance with the requirements of paragraphs 4.4.8.29 and 5.1.2.2.11 of the LSA Code.

Luminous intensity

13.4.1 The luminous intensity of searchlights should be at least 2.5×10^3 candela.

The axial luminous intensity should be at least 90% of the maximum luminous intensity.

The luminous intensity of searchlights should be at a maximum in the centre of the luminous distribution. A homogenous luminous intensity distribution should be ensured.

The effective light emission sectors should be circular and reach vertically and horizontally at least 6°.

Operation time

13.4.2 Searchlights should be suitable for continuous operation of not less than 3 h. During this period the requirements of 13.4.1 above should be fulfilled.

Part 2
Production and installation tests

1 General

1.1 Except where all appliances of a particular type are required by chapter III of the International Convention for the Safety of Life at Sea, 1974, as amended, or the International Life-Saving Appliance (LSA) Code, to be inspected, representatives of the Administration should make random inspection of manufacturers to ensure that the quality of life-saving appliances and the materials used comply with the specification of the approved prototype life-saving appliance.

1.2 Manufacturers should be required to institute a quality control procedure to ensure that life-saving appliances are produced to the same standard as the prototype life-saving appliance approved by the Administration and to keep records of any production tests carried out in accordance with the Administration's instructions.

1.3 Where the proper operation of life-saving appliances is dependent on their correct installation in ships, the Administration should require installation tests to ensure that the appliances have been correctly fitted in a ship.

2 Individual buoyancy equipment

2.1 Lifejackets

Production tests

2.1.1 Manufacturers should be required to carry out a buoyancy test on at least 0.5% of each batch of lifejackets produced, subject to a minimum of one from every batch.

Inspections by the Administration

2.1.2 Inspections by a representative of the Administration should be made at intervals of at least one per 6,000 lifejackets produced, subject to a minimum of one inspection per calendar quarter. When the manufacturer's quality control programme results in lifejackets that are consistently free of defects, the rate of inspection may be reduced to one in every 12,000. At least one lifejacket of each type in production should be selected at random by the inspector and subjected to detailed examination including, if necessary, cutting open. He should also satisfy himself that the flotation tests are being conducted satisfactorily; if he is not satisfied, a flotation test should be undertaken.

2.2 Immersion and anti-exposure suits

Every immersion and anti-exposure suit should be tested with a constant air pressure for a period of at least 15 min and checked for leaks using a leak detection fluid. The air pressure should be appropriate to the type of material used in the manufacture of the suit and should never be less than 0.02 bar. All leaks shall be repaired before the suit leaves the factory.

3 Portable buoyancy equipment

3.1 Lifebuoys

Installation tests

The arrangements for quick release of the lifebuoys fitted with self-activated smoke signals and lights on the ship's navigating bridge should be, using a dummy smoke signal if necessary, tested to demonstrate that the lifebuoys and their attachments will drop clear of the ship's side when released.

4 Pyrotechnics

A statistically adequate sample of pyrotechnics from each batch produced should be activated and observed for proper operation. The tests in section 4 of part 1 should be performed once for every 10 batches of signals produced; however, such tests should be conducted at least once every year, but need not be conducted more often than once in every calendar quarter. Where production of a signal is continuous, the tests in section 4 of Part 1 need only be performed once every year if the Administration is satisfied that the quality control procedures being followed together with continuous production methods make more frequent testing unnecessary.

5 Survival craft

5.1 Liferaft operational inflation test

5.1.1 The Administration should, at its discretion, select a completed and operationally packed liferaft at random and carry out an operational inflation test on a smooth dry floor or on water, e.g., a swimming pool, as a check on the packing and inflation.

5.1.2 The actual distribution of liferafts inflated during a period is left to the Administration's discretion so as to achieve an adequate sampling of the entire production. The selection of the inflatable liferaft or liferafts for the test should be on a random basis. Personnel fabricating and packing inflatable liferafts should not be made aware of which liferaft will be tested until after the liferaft has been packed in its container. The painter should be pulled from the liferaft using a device to measure the applied force. The force required to pull the painter and start inflation should not exceed 150 N. The inflatable liferaft should break free from its container and attain its design shape and full erection of the canopy support tubes in not more than 1 min.

5.1.3 Each liferaft produced should be inspected for defects and dimensional deviations.

5.1.4 Each liferaft produced should be inflated with air to the lesser of 2.0 times its working pressure or that sufficient to impose a tensile load on the inflatable tube fabric of at least 20% of the minimum required tensile strength. Relief valves should be inoperative for this test. After 30 min the liferaft should not show any signs of seam slippage or rupture, nor should the pressure decrease by more than 5%. The measurement of the pressure drop due to leakage can be started when it has been assumed that the compartment rubber material has completed stretching due to the inflation pressure and stabilized. This test should be conducted after equilibrium condition has been achieved. Following the test each relief valve should be tested for proper relief and reseating pressure.

5.1.5 The gas-tight integrity of each inflated compartment of each liferaft produced should be checked by inflating with air to its working pressure. After a settling time of 30 min the pressure should be checked and adjusted to the working pressure as necessary. After 1 h the pressure should not have decreased by more than 5% after compensation for temperature and barometric pressure changes. More than one compartment may be tested at one time, but adjacent compartments with common pressure barriers should be open to the atmosphere during the test.

5.1.6 If the insulation of the floor of the liferaft is obtained by inflation, it should be inflated to its designed pressure. After a period of 1 h, the pressure should not have decreased by more than 5% uncorrected pressure change.

5.1.7 Exact NAP-test pressures can be calculated in accordance with the following equation:

$$P(\text{kg/cm}^2) = \frac{2 \times \text{tensile strength (kg per 5 cm)}}{25 \times \text{diameter tube (cm)}}$$

5.2 Davit-launched liferaft and inflated rescue boat test

Every new davit-launched liferaft and inflatable rescue boat should satisfactorily undergo a 10% overload test in accordance with the approved drawings or construction specification before the final inflation pressure test. The conditions of the 10% overload suspension test are:

.1 the liferaft or rescue boat should be inflated preferably with air and stabilized at its working pressure;

.2 the working pressure should be determined by the reseat of the relief valves. The pressure relief valves should be fully operational;

.3 the floor of the inflatable liferaft should not be inflated;

.4 the 10% overload to be 10% of the mass of the liferaft or rescue boat assembly together with its full equipment and complement of persons calculated at 75 kg per person, for the liferaft and 82.5 kg per person for the rescue boat;

.5 the loaded liferaft or rescue boat should remain suspended for not less than 5 min; and

.6 the inflatable liferaft or rescue boat should not sustain damage to its suspension members, their attachments, or any other structural component as a result of this test. The pressure relief valves should maintain the normal working pressure of the buoyancy tubes and their basic shape during suspension.

5.3 Lifeboat and rescue boat test

5.3.1 Each new davit-launched lifeboat and rescue boat should be loaded to 1.1 times its related load and suspended from its release mechanism. The lifeboat or rescue boat should then be released with the load on the release mechanism. It should also be confirmed that the lifeboat or rescue boat will release when fully waterborne in the light condition and in a 10% overload condition.

5.3.2 Each new free-fall lifeboat should be loaded to 1.1 times its related load and launched by free fall with the ship on an even keel and in its lightest seagoing condition.

5.3.3 Each lifeboat and rescue boat should be operated for at least 2 h before it is installed on the ship. The test should include operation of all systems, including operation of the transmission through all of its positions.

5.3.4 The connection of each release gear which is fixed to the boat should be subjected to a load equal to the weight of the boat with its full complement of persons and equipment (or two times the weight of the boat in the case of single fall systems). There should be no damage to the release gear or its connection to the boat.

5.4 Launch test

It should be demonstrated that the fully equipped lifeboat on cargo ships of 20,000 gross tons or more and rescue boat can be launched from a ship proceeding ahead at a speed of not less than 5 knots in calm water and on an even keel. There should be no damage to the lifeboat or the rescue boat or their equipment as a result of this test.

6 Launching and stowage arrangements

6.1 Launching appliances using falls and winches

Factory overload test

6.1.1 Each launching appliance, except the winch, should be tested with a static load of 2.2 times the working load with the appliance in the full outboard position. For a free-fall lifeboat launching appliance, each launching ramp and its connection to the release mechanism should be tested with a static load of 2.2 times the working load. The appliance should not be deformed or damaged. Winches with the brakes applied should be tested by applying a static load of 1.5 times the maximum working load. Any cast components of the frame and arm should be hammer-tested to determine that they are sound and without flaw.

Installation tests

Loaded test

6.1.2 The survival craft or rescue boat, loaded with its normal equipment or an equivalent mass and a distributed mass equivalent to that of the number of persons, each weighing 75 kg or 82.5 kg, it is permitted to accommodate, should be released by operation of the launching control on deck. The speed at which the survival craft or rescue boat is lowered into the water should be not less than that obtained from the formula:

$$S = 0.4 + (0.02H)$$

where:

S = speed of lowering (m/s)
H = height from davit head to the waterline at the lightest seagoing condition (m).

The maximum lowering speed established by the Administration should not be exceeded.

Light loaded test

6.1.3 The survival craft or rescue boat loaded with its normal equipment or an equivalent mass should be released by operation of the launching control on deck to demonstrate that the lifeboat's mass is sufficient to overcome the frictional resistance of the winch, falls, blocks and associated gear. The lowering speed should be as established by the Administration. A person should then board the survival craft or rescue boat and perform a test of the launching operation from within the boat.

6.1.4 The requirements of 6.1.2 and 6.1.3 do not apply to free-fall lifeboats.

Loaded lowering test (brake test only)

6.1.5 The survival craft or rescue boat loaded with its normal equipment or an equivalent mass and a distributed mass equal to that of the number of persons, each weighing 75 kg or 82.5 kg, it is permitted to accommodate +10% of the working load, should be released by the operation of the launching controls on deck. When the craft has reached its maximum lowering speed, the brake should be abruptly applied to demonstrate that the attachments of the davits and winches to the ship's structure are satisfactory. The maximum lowering speed established by the Administration should not be exceeded.

6.1.6 If lowering of the lifeboat is controlled from within the lifeboat by means of a control wire paid off from an auxiliary drum on the winch, the following additional points should receive particular consideration after installation of the davits and winches:

.1 the mass on the control wire should be sufficient to overcome the friction of the various pulleys on the control wire, when turning out the lifeboat from the stowed to the embarkation position;

.2 it should be possible to operate the winch brake from within the lifeboat;

.3 the winch brake should not be affected by the mass of the fully extended control wire;

.4 there should be sufficient length of control wire available at the lifeboat, during all stages of lowering; and

.5 means should be provided to retain the free end of the control wire in the lifeboat until the lifeboat is detached from the launching appliance by the operator.

6.1.7 If the winch brake is exposed to the weather, the lowering test should be repeated with the braking surface wetted.

Recovery test

6.1.8 It should be demonstrated that the davit-launched lifeboat or rescue boat can be recovered to its stowage position by means of operating the hand gear and can be safely and properly secured.

6.1.9 For free-fall lifeboats it should be demonstrated that the survival craft can be recovered to its stowage position and can be safely and properly secured.

6.1.10 Where davits are recovered by power, it should be demonstrated that the power is automatically cut off before the davit arms come against the stops.

6.1.11 In the case of rescue boat launching appliances, it should be demonstrated that the fully equipped rescue boat when loaded with a mass equal to that of the number of persons it is approved to carry can be recovered by means of a winch at a rate of no less than 0.3 m/s.

6.1.12 It should be demonstrated that the rescue boat can be recovered by means of the winch referred to in 6.1.11 using a hand gear.

Adjustable ramp test

6.1.13 It should be demonstrated that adjustable ramps for free-fall launching may be adjusted satisfactorily with the free-fall lifeboat loaded to 1.2 times its related load.

6.2 Installation tests of liferaft launching appliances

Testing of release arrangements

6.2.1 When the hooks are made of cast steel, acceptable non-destructive tests should be carried out to establish that the material is free from surface or internal flaws.

Static load test

6.2.2 Each release hook should be statically proof-tested to 2.5 times the safe working load and be provided with an approved testing establishment certificate certifying that it has been so tested.

Operational test

6.2.3 Each release hook should be submitted to an operational test with a mass equivalent to the safe working load being applied. The release arrangements should be demonstrated and checked with the liferaft loaded to ensure that the automatic release hook will not release while the load is still applied.

Marking

6.2.4 Each release hook should be checked to ensure it is permanently marked with:

.1 the manufacturer's name or the approved name of the release hook;

.2 the date of manufacture;

.3 the safe working load;

.4 the number of the test certificate required by 6.2.2; and

.5 clear, concise operating instructions.

Lowering test

6.2.5 One liferaft ballasted to represent a 10% overload or an equivalent mass should be lowered from each launching appliance to establish the rate of lowering. The 10% overload should be 10% of the mass of the liferaft assembly together with its equipment and full complement of persons calculated at 75 kg per person. It should be jerked to ensure that the liferaft launching appliance, its fastenings and the supporting structures can withstand the associated loads.

Recording of lowering test

6.2.6 The time should be recorded for the sequence of preparing, loading and launching three liferafts. If so desired, persons may be used only in the preparing and loading operations and ballast substituted for the lowering and launching part of the test. This sequence test need not be carried out on every launching appliance on a ship. However, at least one example of each launching appliance type and arrangement should be so tested on each ship.

Towing strain test

6.2.7 A moderate towing strain should be put on the liferaft when waterborne to check that the release arrangements are satisfactory under this condition.

7 Marine evacuation systems

7.1 Installation tests

7.1.1 On the installation of a marine evacuation system on a ship, at least 50% of such systems should be subjected to a harbour trial deployment. At least one of these systems should be deployed in association with at least two of the inflatable liferafts to establish that correct launching and subsequent retrieving, bowsing-in and inflation procedures have been correctly installed.

7.1.2 Subject to the above deployments being satisfactory, untried systems should be similarly deployed within 12 months of the installation date.

7.1.3 For the first of the above deployments, in association with the launching of the liferafts, a partial evacuation trial should be carried out to ensure that:

- **.1** the system does not interfere with the launching of other life-saving equipment fitted on board; and
- **.2** the system and associated liferafts are clear of all possible obstructions or dangers such as stabilizers or the ship's propellers.

Annex 1

Adult reference test device (RTD) design and construction

1 General

The RTD is intended for use only as a test reference standard to represent the desired level of in-water performance of a lifejacket required by the 1974 SOLAS Convention, and is not considered representative of any other required lifejacket performance. The adult RTD is designed to fit persons from a chest size of 700 mm to 1,350 mm and to be comfortable to wear as a non-reversible device such that it would be obvious to the wearer as to which is the inside and outside of the device, even under reduced lighting conditions. The adult RTD is made with two types of buoyant foam in a vest style using a heavy nylon cover fabric shell secured to the body with 25 mm webbing, closures and adjustments. The shell is made with slide fasteners (zippers) in place of closing seams to hold the foam within, in order that the foam inserts can be easily removed to check their buoyancy and renew or supplement them if they are out of tolerance. Hook and loop fasteners are used on the interior foam retainers to position and prevent shifting of the foam panels.

2 Materials

All materials used should comply with ISO 12402-7.

2.1 Foam requirements

The performance of the RTD is dependent on using plastic foam of the proper stiffness, shape and buoyancy.

2.1.1 Stiffness

Two different stiffness foams are used: one is a soft foam and the other is a stiff foam. A bridge deflection test is provided to determine acceptability for the intended application. Figure A.1 provides the setup details and table A.1 provides the specific measured values. For selecting the type of foam for the specific insert, see tables A.2 and A.3. To measure the centre deflection of a foam panel of the specified cross-section (a × b) and 110 mm wide, place the foam panel centred across the two equal height, parallel horizontal surfaces separated by the specified distance (c), and then load with a mass of the specified width. Note the length of the load should be at least 110 mm, such that when placed on the foam panel it will extend the full width of the foam panel. It is acceptable for the load to extend beyond the width of the foam panel provided that it is centred over the panel with equal amounts extending over the sides of the foam panel. Measure the deflection at the bottom centre location of the foam panel 30 s after placing the load on the panel.

2.1.2 Shape

The shape of each foam insert is specified in figures A.8 to A.11. For dimensions see tables A.2 and A.4.

2.1.3 Buoyancy

The total design buoyancy of the device is 149 N. Table A.3 specifies the foam characteristics, the buoyancy for each insert and its tolerances and the overall buoyancy distribution to be verified when using the RTD for certification testing.

2.2 Other component requirements

See table A.2.

3 Construction

The construction and assembly of the device should be in accordance with tables A.2 to A.4 and figures A.2 to A.14. A tolerance of ± 6 mm is used throughout for fabric cutting and stitching assembly. A tolerance of ± 6 mm is also used for foam cutting, however, the buoyancy requirements of table A.3 should be met.

3.1 Seams

The seam allowances are 13 mm, unless otherwise specified. All structural seams use a lock type stitch so that the seam will not unravel when a force is applied in the direction of the seam on any of the threads forming the stitch. Stitching should have a density of 7 to 12 stitches per 25 mm of stitch length. The box-x stitching on the webbing is 15 mm × 18 mm, unless otherwise specified. The bar-tack stitching on the webbing is 15 mm × 2 mm.

3.1.1 On the closing seam of the back section of the outer and inside cover, the cut ends of the fabric are turned under and stitched so that the fabric will not ravel. The cut ends of webbing should be heat-sealed.

3.1.2 Tabs on the ends of the waist belt are formed by turning under 40 mm of material twice and stitching 19 mm from the end of the folds with box-x or bar tack stitching.

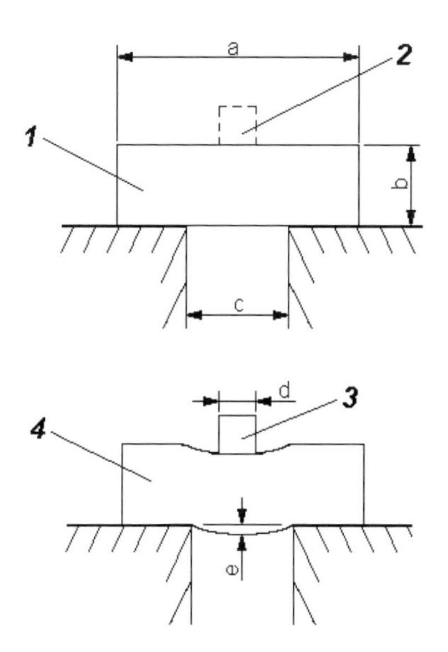

Key

1	Foam at initial setup
2	Centre load
3	Load
4	Foam bridge deflection after 30 seconds

Figure A.1 – *Foam bridge deflection test*

Table A.1 – *Specifications for the foam bridge deflection test*

Foam type	Dimension shown in figure A.1						Load mass kg
	a (Length) mm	(Not shown) (width) mm	b (Thickness) mm	c (Span) mm	d (Load width) mm	e (Deflection) mm	
Stiff	394	110	83	300	120	< 20	8,6
Soft	394	110	45	150	30	≥ 25	0,75

Table A.2 – *Parts, quantity and assembly*

1 Cover fabric	420 denier nylon, with ravel resistant coating, orange			
1.1 Front outer cover		1	A.2	
1.2 Back outer cover		1	A.2	
1.3 Inside cover		1	A.3	
1.4 Centre gusset		2	A.4	
1.5 Collar, outer and inside cover		2	A.5	
1.6 Fabric reinforcement		4	A.6 A.14	Attach to inside of collar cover, as attachment 1, for reinforcement at webbing attachment (see figure A.14).
1.7 Interior fabric retainers for foam inserts 1		2	A.7 A.13	Attach to inside of front cover, as attachment 3, stitch to cover at each side to form a foam retainer for inside front foam insert components 2.2.1 and 2.2.2 (see figure A.13).

Table A.2 *(cont'd)*

1.8 Interior fabric retainers for foam inserts 2		2	A.7 A.14	Attach hoop and loop fasteners to the ends and stitch at centre to the inside of front cover, as attachment 4, to form a foam retainer for front foam insert components 2.1.1 and 2.1.2 (see figure A.13).
2 Foam				
2.1 Stiff	See tables A.1 and A.3			
2.1.1 Front foam insert, right side	81 mm thick	1	A.8	
2.1.2 Front foam insert, left side	81 mm thick	1	A.8	
2.1.3 Collar foam insert	56 mm thick	1	A.10	
2.2 Soft	See tables A.1 and A.3			
2.2.1 Inside front foam insert, right side	46 mm thick	1	A.9	
2.2.2 Inside front foam insert, left side	46 mm thick	1	A.9	
2.2.3 Back foam insert	32 mm thick	1	A.11	

Table A.2 *(cont'd)*

3 Webbing	25 mm, polypropylene, with easy adjustment and no significant slippage when used with the specified hardware.			
3.1 Chest strap	127 mm, black	2	A.12	On left side of front cover, attach webbing with male buckle. On right side of front cover attach webbing with female buckle. The free ends of the chest strap are folded under the yellow webbing (collar attachment webbing), with reinforcing fabric (see figure A.6) on inside of cover fabric. A box-x stitch is used to attach the chest strap to the front cover.
3.2 Waist belt	152 mm, black	2	A.12	On left side attach waist belt with slide and buckle clip waist belt. On right side attach bottom belt with D-ring and slide.

Table A.2 *(cont'd)*

3.3 Waist belt	1,867 mm, black	1	A.12 A.13	Form 40 mm tab on each end. Attach to back cover using three box-x stitches (after front and back covers are assembled).
3.4 Belt loop on front cover	76 mm, black	2	A.12	Attach webbing to front outer cover and form a belt loop (one on each side) by two sets of double bar tack stitches
3.5 Belt loop on inside cover	89 mm, black	2	A.13	Attach webbing to inside cover and form a belt loop (one on each side) by two box-x stitches
3.6 Collar attachment	1,384 mm, yellow	1	A.14 A.6 A.12	Attach webbing to collar and reinforcing fabric, in two places using box-x stitch
4 Hook and loop fastener	50 mm × 70 mm, black generic	2	A.13 A.7	Hook and loop fasteners are attached to the ends of interior fabric retainer for foam insert
5 Thread	Generic synthetic	AR		

Table A.2 *(cont'd)*

6 Hardware				
6.1 Buckle	Male and female 25 mm, plastic, 890 N single-end strength	1		Chest strap
6.2 Slide	Adjuster 25 mm, plastic, 1,600 N single-end strength	2		Waist belt
6.3 Snap hook	25 mm, SS, 1,600 N single-end strength	1		Waist belt
6.4 D-ring	25 mm, SS, 1,600 N single-end strength	2		Waist belt
6.5 Zipper	280 mm, plastic (zipper chain and pulls)	1	A.14	Foam access for collar cover
6.6 Zipper	370 mm, plastic (zipper chain and pulls)	1	A.12	Foam access for back cover
6.7 Zipper	440 mm, plastic (zipper chain and pulls)	2	A.12 A.13	Foam access for front cover

Table A.3 – *Foam insert specifications*
(Values in Newtons (N))

	Front right	Front left	Inside front right	Inside front left	Back	Collar
Foam type*	Stiff	Stiff	Soft	Soft	Soft	Stiff
Buoyancy†	34 ± 1.2	34 ± 1.2	17.5 ± 0.65	17.5 ± 0.65	18 ± 0.8	28 ± 1

* The buoyancy of most foams will change over time with the greatest change occurring in the first several months after manufacture. The exact kind of foam selected will need to be evaluated to determine the amount of additional buoyancy needed at the time of manufacture to maintain the values specified.

† Buoyancy distribution: 69 % front ± 1.5 percentage points.

Table A.4 – *List of dimensions shown in figures A.2 to A.14*
(Dimensions in millimetres)

Letter	Figure											
	A.2	A.3	A.4	A.5	A.6, A.7	A.8	A.9	A.10	A.11	A.12	A.13	A.14
a	72	294	23	308	73	198	76	20	188	100	100	25
b	298	100	516	142	73	46	46	56	274	35	35	160
c	438	1,106	618	10	130	76	394	51	414	20	20	53
d	442	199	102	288	205	81	38	216	343	35	300	25
e	432	398		342	72	76	51	229	147	120	30	45
f	141	597		476	470	157	165	259	223	260		
g	100	1,124		65		394		45		85		
R								70				
h	705	141				46				40		
i	199					8				55		
j	398					20				225		
k	188					20				75		
l	723					76						
m						46						
n						38						
o						165						
p						25						

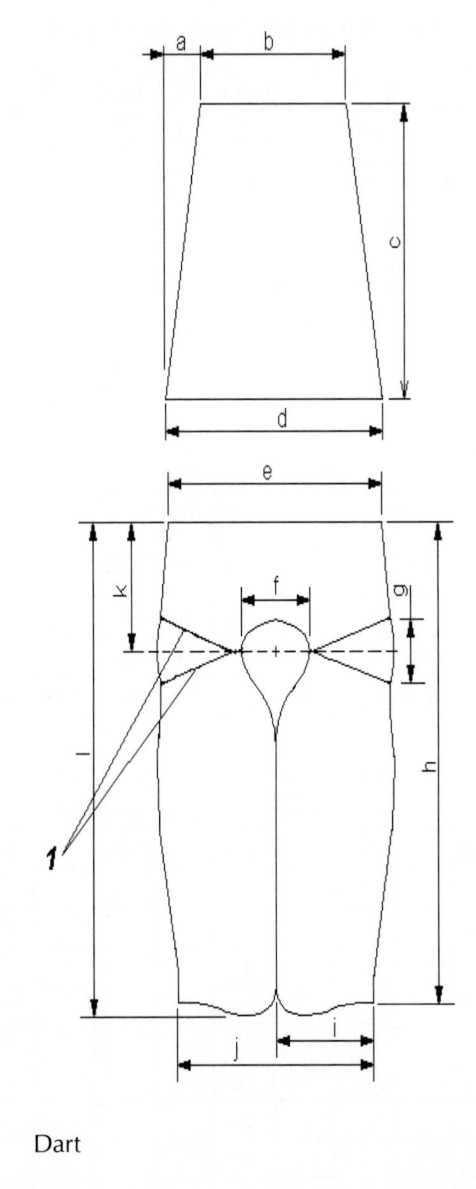

Key
1 Dart

Figure A.2 – *Outer cover, front and back sections*

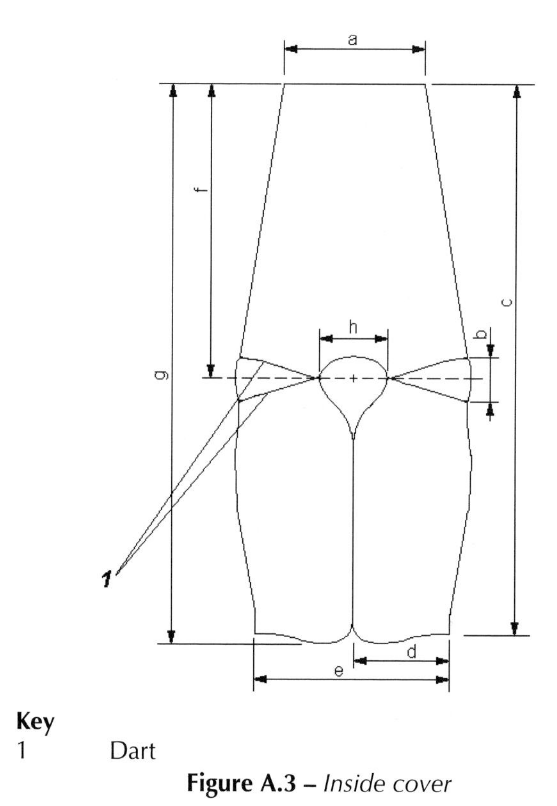

Key

1 Dart

Figure A.3 – *Inside cover*

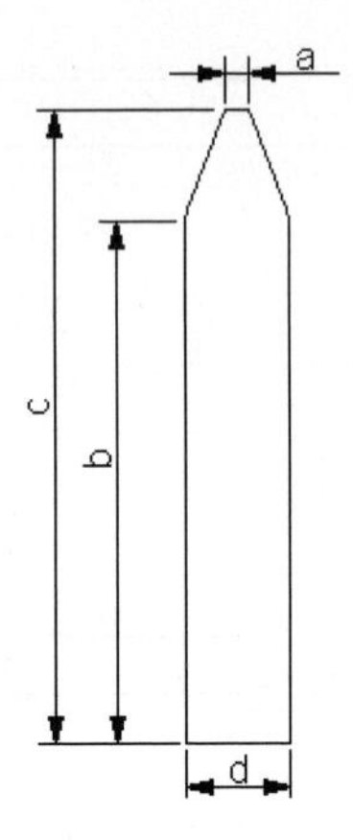

Figure A.4 – *Centre gusset*

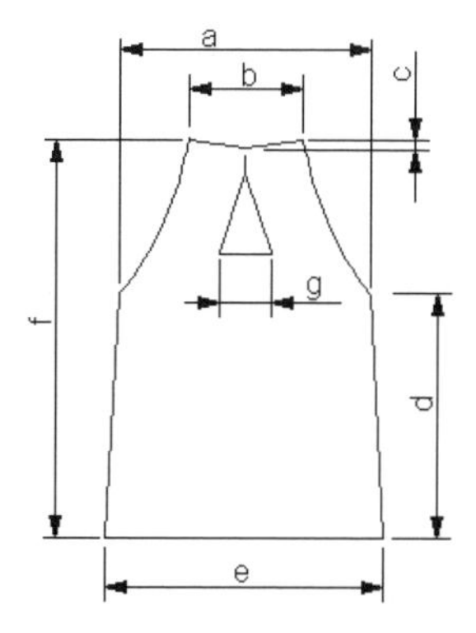

Figure A.5 – *Outer and inside cover, collar*

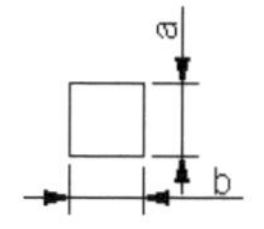

Figure A.6 – *Fabric reinforcement*

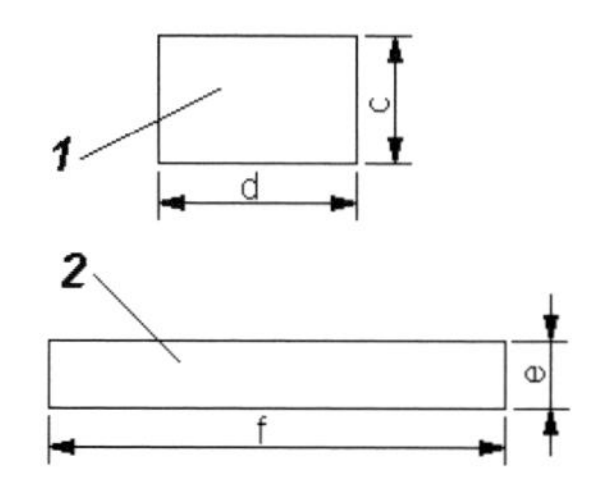

Figure A.7 – *Interior foam retainer*

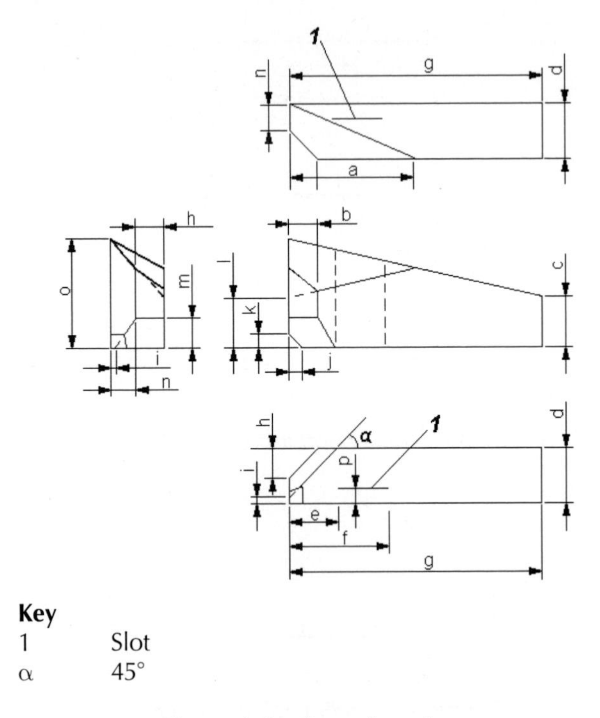

Key

1 Slot

α 45°

Figure A.8 – *Front foam insert*

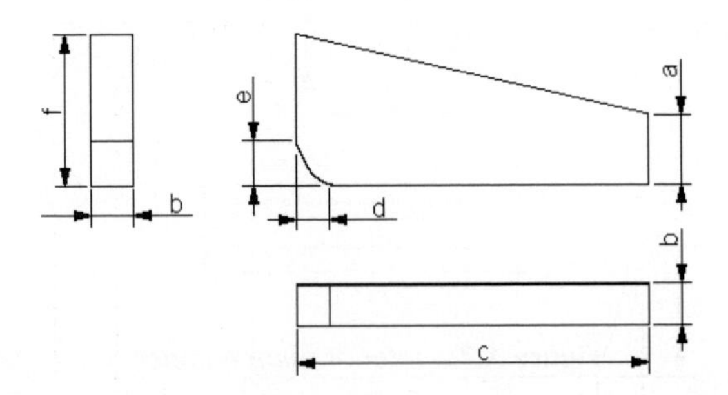

Figure A.9 – *Inside front foam insert*

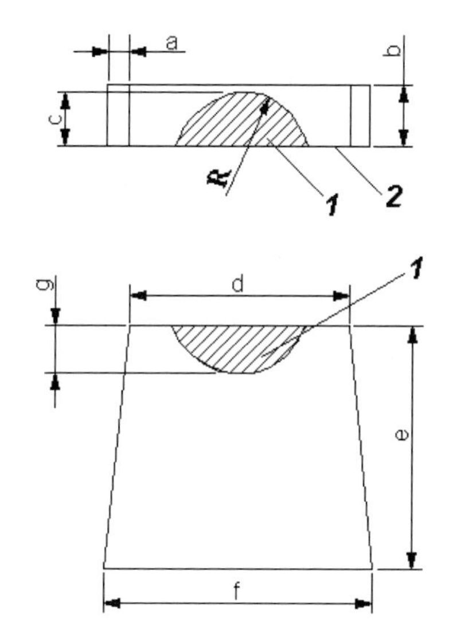

Key

1	Skive
2	Side towards body

Figure A.10 – *Collar foam insert*

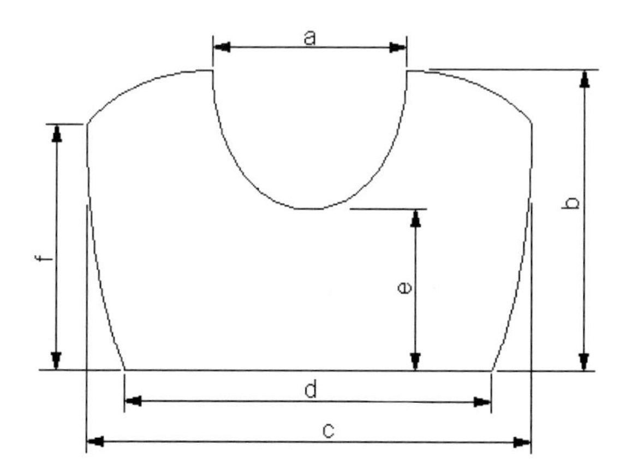

Thickness = 25 mm

Figure A.11 – *Back foam insert*

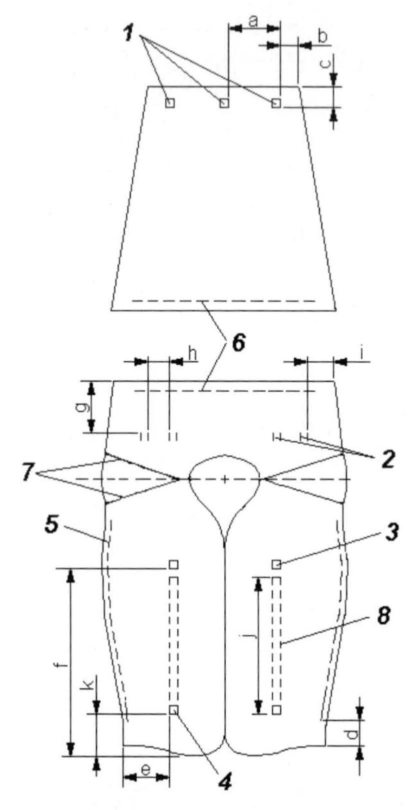

Key

1	Waist belt (1,867 mm) attachment to outside of back cover
2	Zipper (440 mm) attachment to front
3	Chest strap webbing (127 mm) attachment to outside of front cover
4	Waist belt (152 mm) attachment to outside of front cover
5	Belt loop webbing (76 mm) attachment to outside of front cover
6	Zipper (370 mm) attachment to the front and back covers
7	Dart
8	Collar webbing (1,384 mm) attachment to outside of front cover

Figure A.12 – *Attachments to front and back cover*

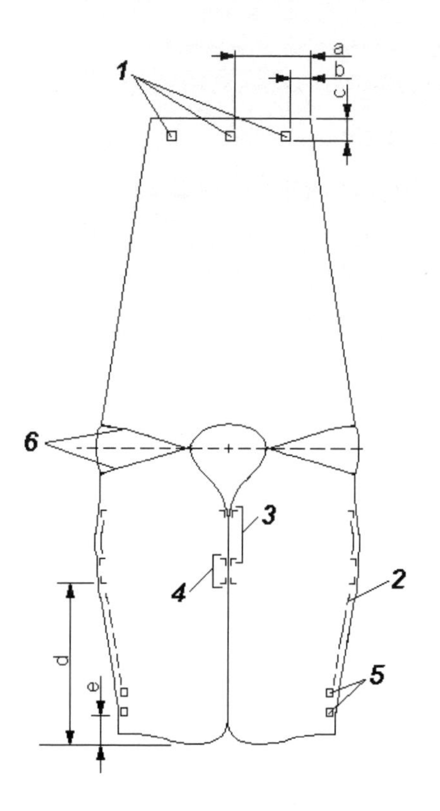

Key

1	Waist belt (1,867 mm) attachment to outside of back cover and inside cover (see figure A.12)
2	Zipper (440 mm) attachment
3	Interior fabric retainer attachment to inside front cover
4	Interior fabric retainer attachment to centre of inside front cover
5	Belt loop webbing (89 mm) attachment to outside of cover
6	Dart

Figure A.13 – *Attachments to inside cover*

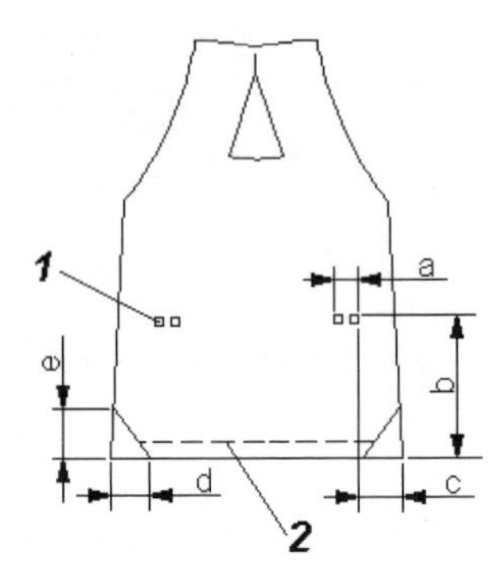

Key

1 Collar webbing (1,384 mm) attachment on the outside of the inner cover with reinforcement fabric inside

2 Zipper (280 mm) attachment to the outer and inner covers

Figure A.14 – *Attachments to outer and inside collar cover*

Appendix

RTD Serial number: ________

ADULT REFERENCE TEST DEVICE – BUOYANCY TRACKING AND VERIFICATION

To achieve repeatability in human subject testing, the overall buoyancy and distribution of buoyancy between the front and back of the RTD must be maintained within a tight tolerance as specified in Table 1.

Table 1 – *SOLAS adult RTD buoyancy and tolerance*

Limit/Units	Front buoyancy*	Back buoyancy	Total buoyancy	Buoyancy distribution†
Design/N	103	46	149	69% in front
Maximum/N	107	48	155	70.5% in front
Minimum/N	100	45	145	67.5% in front

* Values at or corrected to standard temperature and pressure.

† Buoyancy distribution is calculated by dividing the front buoyancy by the total buoyancy.

The buoyancy of a new RTD may exceed the allowable tolerance range until the normal shrinkage or compression of the foam inserts stabilizes. Until the buoyancies of the foam inserts have stabilized, buoyancy and distribution should be checked at regular intervals (perhaps weekly), and then at least monthly thereafter or whenever used for testing, whichever is longer (frequent use may require more frequent checks). Only RTDs with buoyancies within tolerance should be used for certification testing. A data sheet to document the buoyancy and buoyancy distribution of the RTD is attached.

Adjustment of buoyancy: At the time of manufacture the left-to-right distribution of buoyancy in the front inserts was adjusted to be within 1.3 N of each other. To achieve this tolerance, thin layers of foam ("make-up" inserts) may have been inserted between the front and inside front foam inserts. The test house may need to increase the size of these make-up inserts from time to time to keep these parameters within tolerance, or may need to add buoyancy to the back or collar inserts (or trim buoyancy, if the back insert has not shrunk as anticipated). Figure 2 provides guidance for sizing of make-up inserts to adjust buoyancy. After a full sheet of 6.5 mm thick foam is required in any one of the four major areas, an inside front or back insert probably needs to be replaced. If the front buoyancy is under the minimum value, measure the buoyancy of the right and left sides so that the proper distribution of buoyancy (no more than a 1.3 N difference) between the right and left front panels can be maintained.

Table 2 – *SOLAS adult RTD insert design buoyancies*

	Combined left front and inside front*	**Combined right front and inside front***	**Back**	**Collar**
Design (N)	34 + 17.5 = 51.5	34 + 17.5 = 51.5	18	28
S/N ______ Date:				

* Plus make-up inserts, if used.

RTD BUOYANCY DATA SHEET

RTD serial number/identification: ____________

Date	Left front buoyancy (N)*	Right front buoyancy (N)	Total front buoyancy (N)†	Total back buoyancy (N)	Total buoyancy (N)	Buoyancy distribution (% in front)	Remarks

* Left and right front buoyancy need not be checked if distribution is within tolerance.

† If the temperature and pressure at the time of measurement is not at standard conditions, these values should be corrected to standard temperature and pressure.

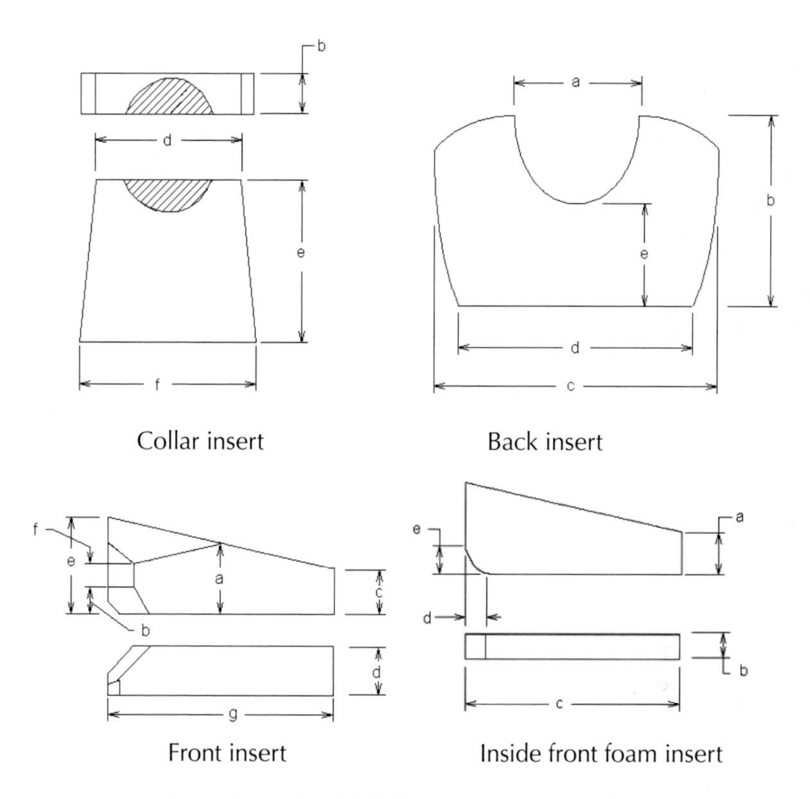

Figure 1 – *Adult RTD foam insert nomenclature*

Buoyancy (N)	Length (mm)	Height (mm)
0.9	84	146
1.3	126	137
1.8	172	126
2.2	222	114
3.1	394	76

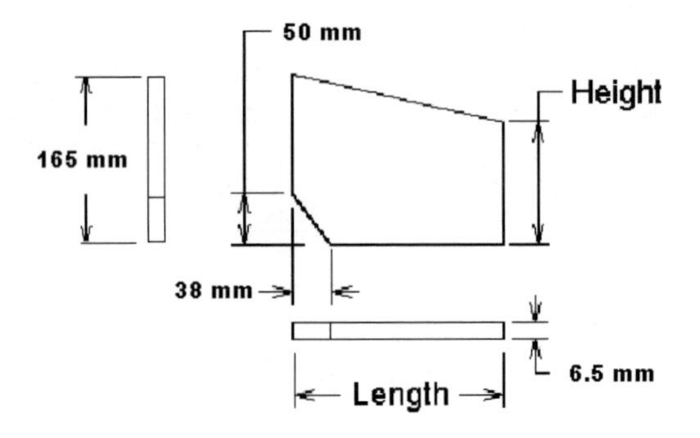

Figure 2 – *Adult RTD "make-up" foam insert sizes*

Annex 2

Child reference test device (RTD) design and construction

1 General

The RTD is intended for use only as a test reference standard to represent the desired level of in-water performance of a lifejacket required by the 1974 SOLAS Convention, and is not considered representative of any other required lifejacket performance. The child RTD is for persons weighing approximately 15 to 43 kg, or 100 to 155 cm in height. The device is designed to fit persons with chest sizes from 50 cm to 70 cm. This RTD is made with layers of buoyant foam in a bib-style design using a heavy nylon shell cover fabric secured to the body with a waist belt with quick and positive closure and adjustment, along with a chest strap at the neck for closure and adjustment. The shell is made with slide fasteners (zippers) in place of closing seams to hold the foam within, in order that the foam inserts can be easily removed to check their buoyancy and renew or supplement them if they are out of tolerance. The RTD is designed to be reasonably comfortable to wear as a non-reversible device.

2 Materials

All materials used should comply with ISO 12402-7.

2.1 Foam requirements

The performance of the RTD is dependent on using plastic foam of the proper stiffness, shape and buoyancy.

2.1.1 Stiffness

The buoyant inserts are made of layers of medium stiffness foam to create a flexible but firm buoyancy element.

2.1.2 Shape

The shape of each foam layer is identified in figures B.2 and B.3. Dimensions are in tables B.1, B.2 and B.3.

2.1.3 Buoyancy

The total design buoyancy of the device is 88 N. Table B.4 specifies foam characteristics, the buoyancy for each insert and its tolerances, and the overall buoyancy distribution to be verified when using the RTD for certification testing.

2.2 Other component requirements

See table B.1.

3 Construction

The construction and assembly of the device should be in accordance with tables B.1 and B.5 and figures B.1 through B.9. A tolerance of ± 6 mm is used throughout for fabric cutting and stitching assembly. A tolerance of ± 6 mm is also used for foam cutting, however, the buoyancy requirements of table A.3 should be met.

3.1 Seams

Seam allowances are 13 mm, unless otherwise specified. All structural seams use a lock type stitch so that the seam will not unravel when a force is applied in the direction of the seam on any of the threads forming the stitch. Stitching should have a density of 7 to 12 stitches per 25 mm of stitch length. Box-x stitching on the webbing is 30 × 15 mm for the waist belt and 15 × 13 mm for the belt loop and chest strap, unless otherwise specified. The bar-tack stitching on webbing is 30 × 2 mm for the waist belt and 15 × 2 mm for the belt loop and chest strap.

3.1.1 The fabric reinforcements for the waist belt, belt loop and chest strap should be attached to the inside surface of the outside cover before attaching any of these items. On the closing seam of the top and bottom sections of the outside and inside cover, the cut ends of the fabric are turned under and stitched when installing the zippers so that the fabric will not fray and so that the folds are flush with the line where the zipper teeth mesh (zippers installed to be hidden by cover fabric when closed).

Table B.1 – *Parts, quantity and assembly*

1	**Cover fabric**	420 denier nylon, with ravel resistant coating, orange		B.1, B.4, and B.9	
1.1	Outside cover		1	B.1, B.4, and B.9	
1.2	Inside cover		1	B.1, B.4, and B.9	
1.3	Fabric reinforcement, chest strap)		2	B.5 and B.9	Attach one each to inside left and right outside covers for the chest strap. Use lock stitches on three sides each (see figure B.9 for locations).
1.4	Fabric reinforcement, belt, and belt loop		2	B.5 and B.9	Attach to inside left and right outside covers for the waist belt and belt loop. Use lock stitches on three sides (see figure B.9 for location).
2	**Foam**	7 mm thickness, polyethylene (PE) foam, except for one layer as needed to achieve required buoyancy		B.2 and B.3	Layers stacked per figures B.2 and B.3.
2.1	Front foam insert, left		13 layers	B.2	Trim corner of layers A and B only per figure B.2.
2.2	Front foam insert, right		13 layers	B.2	Trim corner of layers A and B only per figure B.2.
2.3	Back foam insert		11 layers	B.3	

Table B.1 *(cont'd)*

Component	Description	Quantity	See figure	Construction notes
3 Webbing				All cut ends heat-sealed.
3.1 Waist belt webbing	38 mm, black, polypropylene, with easy adjustment and no significant slippage when used with the specified hardware.	1,285 mm cut length	B.1, B.8 and B.9	On left side attach waist belt with fixed part of buckle. Tab on the end of belt formed by turning under 40 mm of material twice and stitching 19 mm from the end of the fold with a bar-tack stitch. For location see figure B.9.
3.2 Belt loop webbing	19 mm, black, polypropylene.	80 mm cut length	B.1 and B.9	Attach webbing to front outside cover with two sets of double bar tack stitches to form a belt loop. For location see figure B.9.
3.3 Chest strap webbing	19 mm, black, polypropylene.	235 mm and 80 mm cut length	B.1, B.7 and B.9	Attach webbing with female buckle to right outside cover. Attach webbing with male buckle to left outside cover. For location see figure B.9. Tab formed 75 mm from the free end of the male section of chest strap by folding in "Z" pattern 30 mm apart and stitching 15 mm from the fold with a bar-tack stitch. See figure B.7.

Table B.1 *(cont'd)*

Component	Description	Quantity	See figure	Construction notes
4 Thread	Generic synthetic	AR		
5 Hardware				
5.1 Buckle	38 mm, plastic (male and female sections)	1	B.1 and B.8	Used in waist belt assembly
5.2 Buckle	19 mm, plastic (male and female sections)	1	B.1 and B.7	Used in chest strap assembly
5.3 Zipper	380 mm, plastic (zipper chain length)	1	B.1 and B.9	
5.4 Zipper	150 mm, plastic separating (zipper chain and box/pin length)	2	B.1 and B.9	

Table B.2 – *List of dimensions shown in figure B.2*

Dimension	Insert layer dimensions (mm)				
	A	***B***	***C***	***D***	***E***
a	145	140	125	115	95
b	305	300	285	275	255
c	30	30	0	0	0
d	30	30	0	0	0

Table B.3 – *List of dimensions shown in figure B.3*

Dimension	Insert layer dimensions (mm)				
	A	***B***	***C***	***D***	***E***
a	343	335	325	315	305
b	140	133	120	108	95
c	9	5	3	0	–5*
R	46	50	52	55	55

* Measured in direction opposite to that shown in figure.

Table B.4 – *Foam insert specifications*

	Left front insert	Right front insert	Back insert
Density	29 ± 5 kg/m^3	29 ± 5 kg/m^3	29 ± 5 kg/m^3
Compressive strength at 25% (ISO 3386-1)	35 ± 10 kPa	35 ± 10 kPa	35 ± 10 kPa
Buoyancy [a, b]	31.5 ± 1.2 N	31.5 ± 1.2 N	25 ± 1.2 N

[a] The buoyancy of most foams will change over time with the greatest change occurring in the first several months after manufacture. The exact kind of foam selected will need to be evaluated to determine the amount of additional buoyancy needed at the time of manufacture to achieve the values specified.

[b] Buoyancy distribution: 71.5% front ± 1.5 percentage points

Table B.5 – *List of dimensions shown in figures B.4–B.9*

Dimension	Dimensions (mm)						
	Figure B.4	Figure B.5		Figure B.6	Figure B.7	Figure B.8	Figure B.9
		Key-1	Key-2				
a	420	75	80	75	90	1,150*	45
b	210	105	110		40		135
c	92						85
d	210						45
e	356						25
f	230						33
g	460						115
h	375						25
i	580						265

* With webbing assembly fully extended.

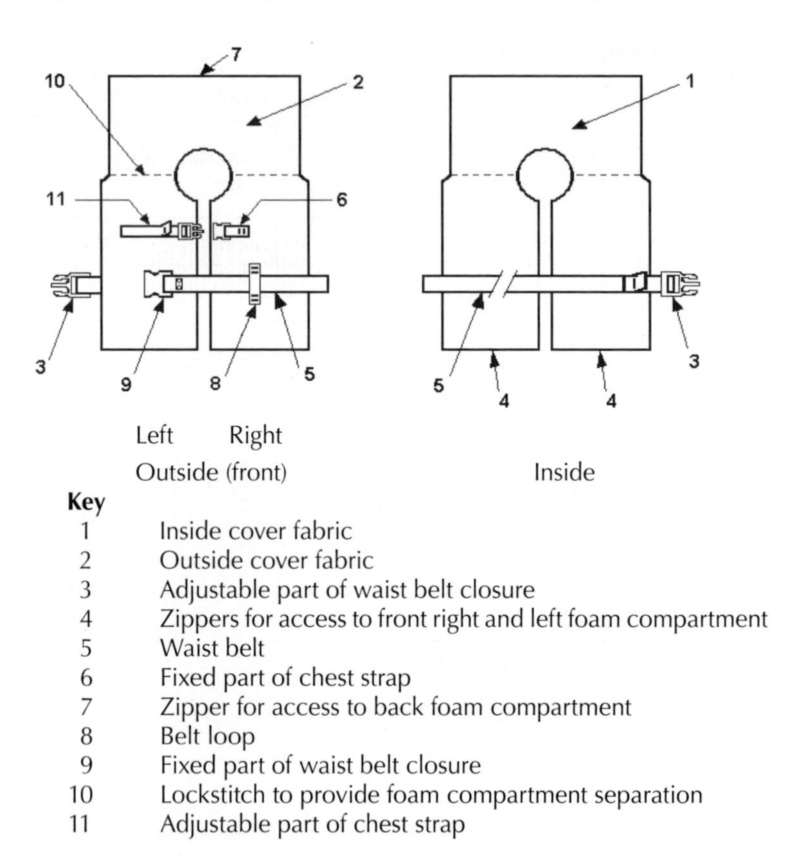

Key

1	Inside cover fabric
2	Outside cover fabric
3	Adjustable part of waist belt closure
4	Zippers for access to front right and left foam compartment
5	Waist belt
6	Fixed part of chest strap
7	Zipper for access to back foam compartment
8	Belt loop
9	Fixed part of waist belt closure
10	Lockstitch to provide foam compartment separation
11	Adjustable part of chest strap

Figure B.1 – *General arrangement, right side out (outside and inside)*

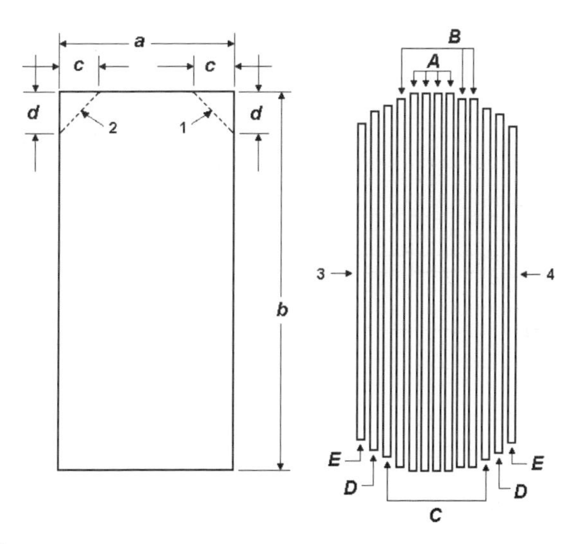

Key

1 Trim upper right corner only for left insert layers per table B.2
2 Trim upper left corner only for right insert layers per table B.2
3 Outside
4 Inside

Figure B.2 – *Front foam inserts (right and left sides)*

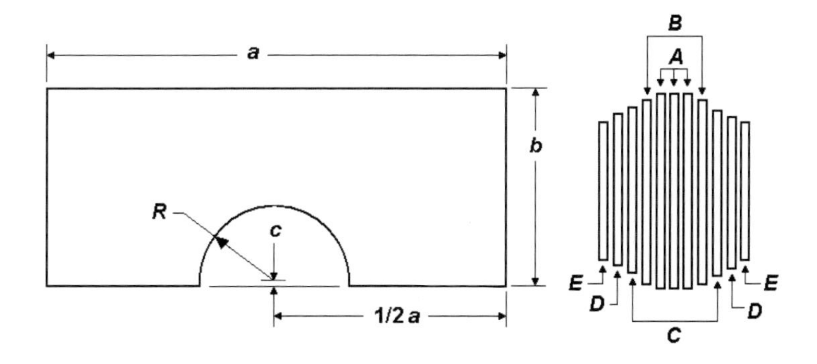

Figure B.3 – *Back foam insert*

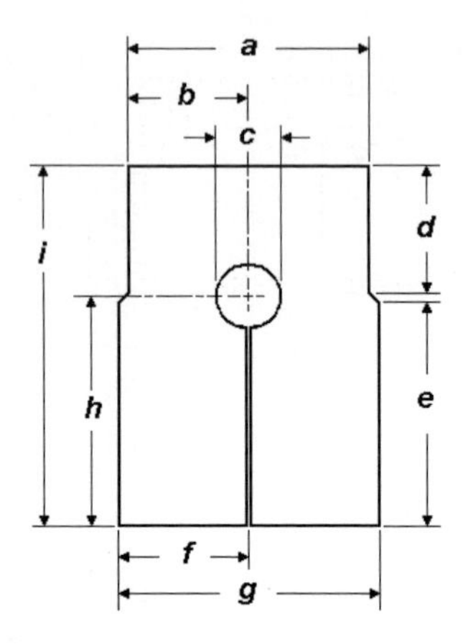

Figure B.4 – *Cover cut pattern (outside and inside covers)*

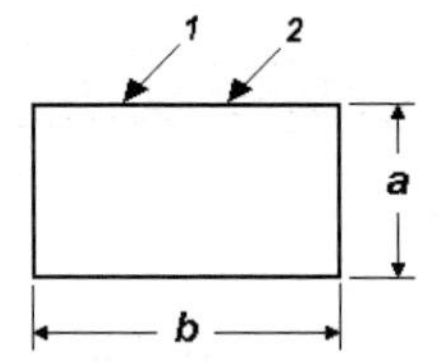

Key

1 Fabric reinforcements for chest strap attachments

2 Fabric reinforcement for waist belt and belt loop attachments

Figure B.5 – *Fabric reinforcements*

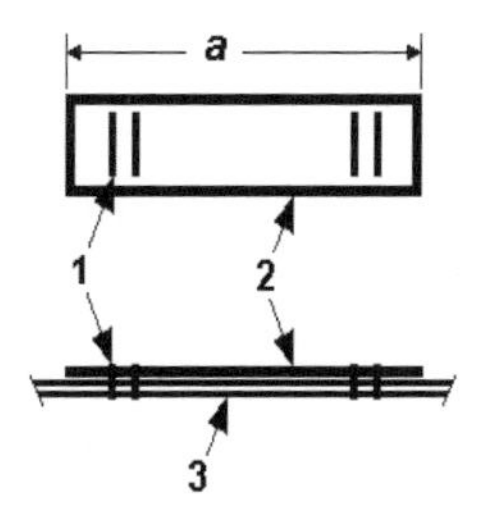

Key

1	Bar-tack
2	Webbing
3	Outer cover and reinforcement (shown on lower view only)

Figure B.6 – *Belt loop*

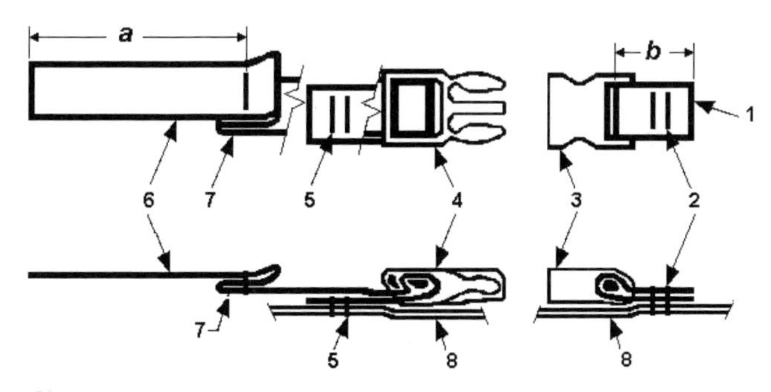

Key

1	Webbing
2	Double bar-tack (or Box-x) stitch
3	Fixed part of closure
4	Adjustable part of closure
5	Double bar-tack (or Box-x) stitch
6	Webbing
7	Tab
8	Outer cover and reinforcement (shown on lower view only)

Figure B.7 – *Chest strap assembly*
(adjustable part, left and fixed part, right)

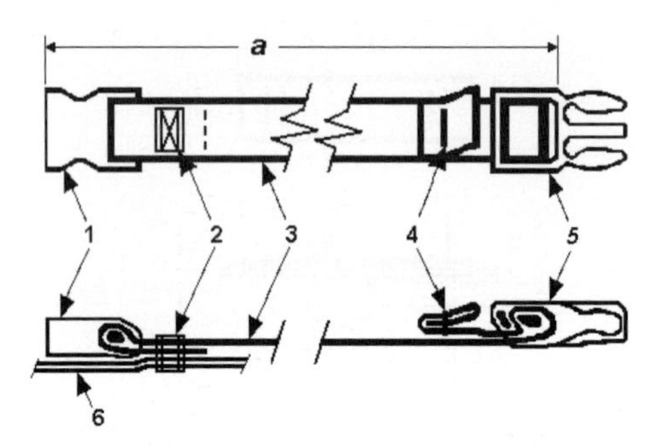

Key

1	Fixed part of closure
2	Box-x (or double bar-tack) stitch
3	Webbing
4	Tab, double fold webbing and secure with a bar-tack stitch
5	Adjustable part of closure
6	Outer left cover and reinforcement (shown on lower view only)

Figure B.8 – *Waist belt assembly*

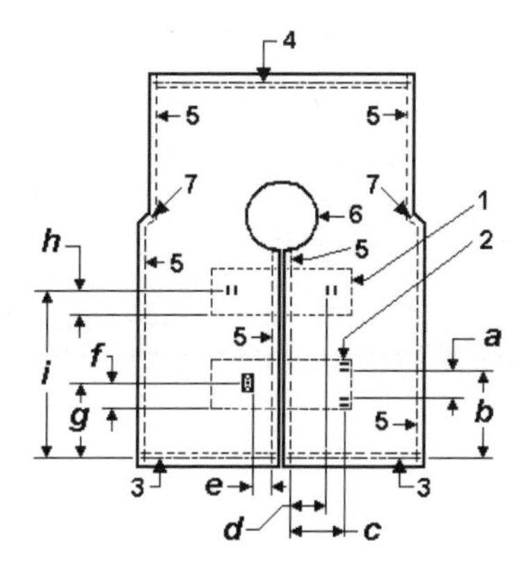

Key

1	Stitching on interior fabric reinforcement for chest strap on right and left sides of the outside cover
2	Stitching on interior fabric reinforcement for waist belt and belt loop on right and left sides of the outside cover
3	Fabric fold and zipper teeth line of engagement when zipper is attached to outside and inside covers
4	Fabric fold and zipper teeth line of engagement when zipper is attached to outside and inside covers
5	Lockstitch seams (with fabric face to face)
6	Lockstitch with 5 mm seam allowance and over-edge stitch (with fabric face to face)
7	After stitching cut relief

Figure B.9 – *Initial assembly (shown right side out, except as noted)*

Appendix

RTD Serial number: _______

CHILD REFERENCE TEST DEVICE – BUOYANCY TRACKING AND VERIFICATION

To achieve repeatability in human subject and manikin testing, the overall buoyancy and distribution of buoyancy between the front and back of the RTD should be maintained within a tight tolerance as specified in table 1.

Table 1 – *SOLAS child RTD buoyancy and tolerance*

Limit/Units	Front buoyancy[1, 2]	Back buoyancy[1]	Total buoyancy[1]	Buoyancy distribution[3]
Design/N	63	25	88	71.5% in front
Maximum/N	65.4	26.2	91.6	73% in front
Minimum/N	60.6	23.8	84.4	70% in front

[1] Values at or corrected to standard temperature and pressure.

[2] The left-to-right buoyancy distribution in the front inserts is to be within 1.3 N of each other.

[3] Buoyancy distribution is calculated by dividing the front buoyancy by the total buoyancy.

The buoyancy of a new RTD may exceed the allowable tolerance range until the normal shrinkage or compression of the foam inserts stabilizes. Until the buoyancies of the foam inserts have stabilized, buoyancy and distribution should be checked at regular intervals (perhaps weekly), and then at least monthly thereafter or whenever used for testing, whichever is longer (frequent use may require more frequent checks). Only RTDs with buoyancies within tolerance should be used for certification testing. A data sheet to document the buoyancy and buoyancy distribution of the RTD is attached.

To check buoyancy tolerances, foam inserts need to be removed from the device. Take care that all trapped air is removed when checking buoyancy and that layers are maintained in their proper sequence when reinstalled (considerable effort would be needed to remove entrapped air if testing the intact device).

Adjustment of buoyancy: At the time of manufacture the left-to-right distribution of buoyancy in the front inserts was adjusted to be within 1.3 N of each other. To achieve this tolerance, the layers were individually selected to achieve the cumulative insert buoyancy. If buoyancy of a new device exceeds the upper limits, one layer per compartment may be altered or replaced to bring the unit into compliance. The test house may need to add make-up layers (see figure 3) from time to time to maintain the front-to-back and side-to-side insert tolerances. If the front buoyancy is under the minimum value, measure the buoyancy of the right and left sides so that the proper distribution of buoyancy (no more than a 1.3 N difference) between the right and left front panels can be maintained.

Table 2 – *SOLAS child RTD insert design buoyancies*

	Combined left front insert (13 layers)	**Combined right front insert (13 layers)**	**Combined back insert (11 layers)**
Design (N)	31.5	31.5	25
S/N ______ Date:			

RTD BUOYANCY DATA SHEET

RTD Serial number/identification: ____________

Date	Left front buoyancy (N)[1]	Right front buoyancy (N)	Total front buoyancy (N)[2]	Total back buoyancy (N)	Total buoyancy (N)	Buoyancy distribution (% in front)	Remarks

[1] Left and right front buoyancy need not be checked if distribution is within tolerance.

[2] If the temperature and pressure at the time of measurement is not at standard conditions, these values should be corrected to standard temperature and pressure.

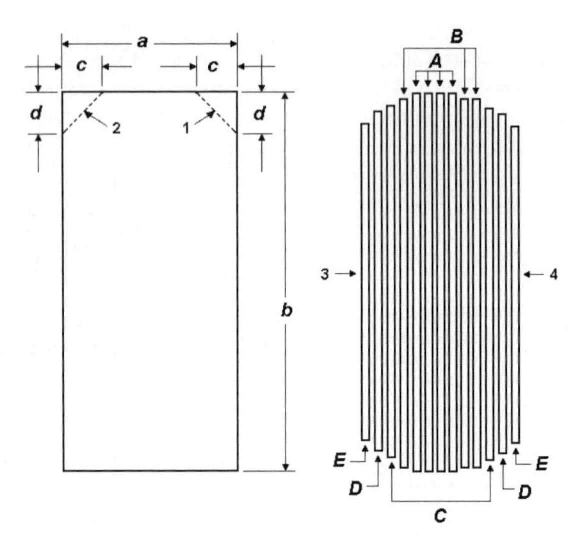

Key

1	Trim upper right corner only for left insert layers A and B.
2	Trim upper left corner only for right insert layers A and B.
3	Outside
4	Inside

Insert layer	Buoyancy (approx.) (N)	Insert layer dimensions (mm)			
		a	*b*	*c*	*d*
A	2.8	145	305	30	30
B	2.7	140	300	30	30
C	2.3	125	285	0	0
D	2.0	115	275	0	0
E	1.6	95	255	0	0

Figure 1 – *Front foam insert specifications*

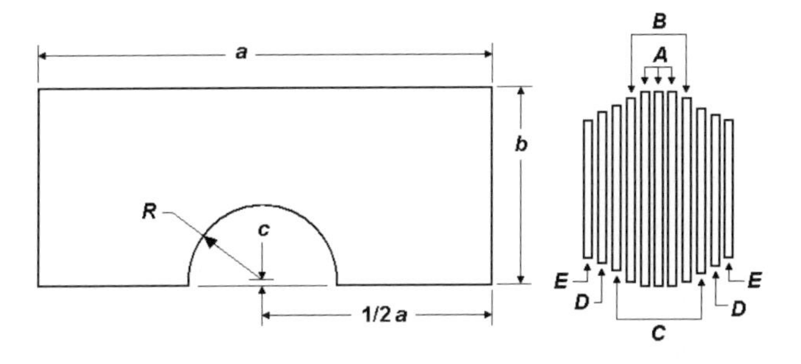

Insert layer	Buoyancy (approx.) (N)	Insert layer dimensions (mm)			
		a	*b*	*c*	*R*
A	2.8	343	140	9	46
B	2.6	335	133	5	50
C	2.2	325	120	3	52
D	1.9	315	108	0	55
E	1.6	305	95	–5*	55

* Measured in direction opposite to that shown in figure.

Figure 2 – *Back foam insert specifications*

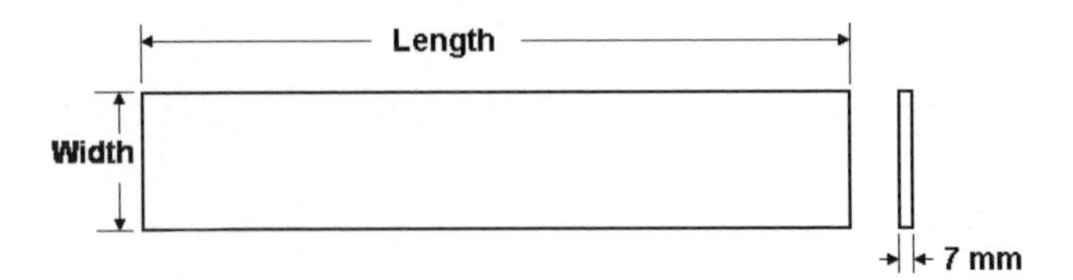

1 Any thickness of foam up to 7 mm is acceptable for a make-up layer.

2 For 7 mm thick foam, 15,300 mm^2 surface area equals approximately 1 N of buoyancy.

Make-up layer[1]	Buoyancy (approx.) (N)	Make-up layer dimensions (mm)	
		Length (mm)[2]	Width (mm)
Front	1.0	300	51
	1.5		76
Back	1.0	340	45
	1.5		67

[1] For 7 mm thick foam.

[2] The length for make-up layer is fixed to maintain proper placement within the lifejacket, but the width may vary to obtain desired buoyancy.

Figure 3 – *Child RTD "make-up" foam insert sizes*

Annex 3

Infant reference test device (RTD) design and construction

1 General

The RTD is intended for use only as a test reference standard to represent the desired level of in-water performance of a lifejacket required by the 1974 SOLAS Convention, and is not considered representative of any other required lifejacket performance. The infant RTD is for persons weighing less than 15 kg, or less than 100 cm in height. The device is designed to fit persons with a chest size of less than 50 cm. This RTD is made with layers of buoyant foam in a bib-style design using a heavy nylon shell cover fabric secured to the body with a waist belt with quick and positive closure and adjustment, along with a chest strap at the neck for closure and adjustment. The shell is made with slide fasteners (zippers) in place of closing seams to hold the foam within, in order that the foam inserts can be easily removed to check their buoyancy and renew or supplement them if they are out of tolerance. The RTD is designed to be reasonably comfortable to wear as a non-reversible device.

2 Materials

All materials used should comply with ISO 12402-7.

2.1 Foam requirements

The performance of the RTD is dependent on using plastic foam of the proper stiffness, shapes, and buoyancy.

2.1.1 Stiffness

The buoyant inserts are made of layers of medium stiffness foam to create a flexible but firm buoyancy element.

2.1.2 Shape

The shape of each foam layer is identified in figures C.2 and C.3. Dimensions are in tables C.1, C.2 and C.3.

2.1.3 Buoyancy

The total design buoyancy of the device is 71 N. Table C.4 identifies foam characteristics, the buoyancy for each insert and its tolerances, and the overall buoyancy distribution to be verified when using the RTD for certification testing.

2.2 Other component requirements

See table C.1.

3 Construction

The construction and assembly of the device should be in accordance with tables C.1 and C.5 and figures C.1 to C.9. A tolerance of ± 6 mm is used throughout for fabric cutting and stitching assembly. A tolerance of ± 6 mm is also used for foam cutting, however, the buoyancy requirements of table C.4 should be met.

3.1 Seams

Seam allowances are 13 mm, unless otherwise specified. All structural seams use a lock type stitch so that the seam will not unravel when a force is applied in the direction of the seam on any of the threads forming the stitch. Stitching should have a density of 7–12 stitches per 25 mm of stitch length. Box-x stitching on the webbing is 30 × 15 mm for the waist belt and 15 × 13 mm for the belt loop and chest strap, unless otherwise specified. The bar-tack stitching on webbing is 30 × 2 mm for the waist belt and 15 × 2 mm for the belt loop and chest strap.

3.1.1 The fabric reinforcements for the waist belt, belt loop and chest strap should be attached to the inside surface of the outside cover before attaching any of these items. On the closing seam of the top and bottom sections of the outside and inside cover, the cut ends of the fabric are turned under and stitched when installing the zippers so that the fabric will not ravel and so that the folds are flush with the line where the zipper teeth mesh (zippers installed to be hidden by cover fabric when closed).

Table C.1 – *Parts, quantity and assembly*

1 Cover fabric	420 denier nylon, with ravel resistant coating, orange		C.1, C.4, and C.9	
1.1 Outside cover		1	C.1, C.4, and C.9	
1.2 Inside cover		1	C.1, C.4, and C.9	
1.3 Fabric reinforcement, chest strap		2	C.5 and C.9	Attach one each to inside left and right outside covers for the chest strap. Use lock stitches on three sides each (see figure C.9 for locations).
1.4 Fabric reinforcement, belt and belt loop		2	C.5 and C.9	Attach to inside left and right outside covers for the waist belt and belt loop. Use lock stitches on three sides (see figure C.9 for location).
2 Foam	7 mm thickness, polyethylene (PE) foam, except for one layer as needed to achieve required buoyancy		C.2 and C.3	Layers stacked per figures C.2 and C.3.
2.1 Front foam insert, left		15 layers	C.2	Trim corners per figure C.2, except outside layers B to G.
2.2 Front foam insert, right		15 layers	C.2	Trim corners per figure C.2, except outside layers B to G.
2.3 Back foam insert		12 layers	C.3	

Table C.1 *(cont'd)*				
Component	**Description**	**Quantity**	**See figure**	**Construction notes**
3 Webbing				All cut ends heat-sealed.
3.1 Waist belt webbing	38 mm, black, polypropylene, with easy adjustment and no significant slippage when used with the specified hardware.	1,085 mm cut length	C.1, C.8 and C.9	On left side attach waist belt with female buckle. Tab on the end of belt formed by turning under 40 mm of material twice and stitching 19 mm from the end of the fold with a bar-tack stitch. For location see figure C.9.
3.2 Belt loop webbing	19 mm, black, polypropylene.	80 mm cut length	C.1, C.6, and C.9	Attach webbing to front outside cover with two sets of double bar tack stitches to form a belt loop. For location see figure C.9.
3.3 Chest strap webbing	19 mm, black, polypropylene.	235 mm and 80 mm cut length	C.1, C.7 and C.9	Attach webbing with female buckle to right outside cover. Attach webbing with male buckle to left outside cover. For location see figure C.9. Tab formed 75 mm from the free end of the male section of chest strap by folding in "Z" pattern 30 mm apart and stitching 15 mm from the fold with a bar-tack stitch. See figure C.7.

Table C.1 *(cont'd)*

Component	Description	Quantity	See figure	Construction notes
4 Thread	Generic synthetic	AR		
5 Hardware				
5.1 Buckle	38 mm, plastic (male and female sections)	1	C.1 and C.8	Used in waist belt assembly.
5.2 Buckle	19 mm, plastic (male and female sections)	1	C.1 and C.7	Used in chest strap assembly.
5.3 Zipper	350 mm, plastic (zipper chain length)	1	C.1 and C.9	Installed to be hidden by cover fabric when closed.
5.4 Zipper	180 mm, plastic separating (zipper chain and box/pin length)	2	C.1 and C.9	Installed to be hidden by cover fabric when closed.

Table C.2 – *List of dimensions shown in figure C.2*

Dimension	Insert layer dimensions (mm)						
	A	***B***	***C***	***D***	***E***	***F***	***G***
a	140	133	127	120	108	95	83
b	190	184	178	172	165	160	140
c	28	28	28	28	28		

Table C.3 – *List of dimensions shown in figure C.3*

Dimension	Insert layer dimensions (mm)				
	A	*B*	*C*	*D*	*E*
a	310	303	290	275	255
b	165	160	140	120	95
c	3	3	3	3	–3*
R	44	44	44	44	44

* Measured in direction opposite to that shown in figure.

Table C.4 – *Foam insert specifications*

	Left front insert	Right front insert	Back insert
Density	29 ± 5 kg/m^3	29 ± 5 kg/m^3	29 ± 5 kg/m^3
Compressive strength at 25% (ISO 3386-1)	35 ± 10 kPa	35 ± 10 kPa	35 ± 10 kPa
Buoyancy [a, b]	21 ± 1.2 N	21 ± 1.2 N	29 ± 1.2 N

[a] The buoyancy of most foams will change over time with the greatest change occurring in the first several months after manufacture. The exact kind of foam selected will need to be evaluated to determine the amount of additional buoyancy needed at the time of manufacture to achieve the values specified.

[b] Buoyancy distribution: 59.2 % front ± 1.5 percentage points

Table C.5 – *List of dimensions shown in figures C.4–C.9*

Dimension	Dimensions (mm)						
	Figure C.4	Figure C.5		Figure C.6	Figure C.7	Figure C.8	Figure C.9
		Key-1	Key-2				
a	390	75	80	75	90	950*	45
b	195	105	110		40		115
c	85						140
d	220						45
e	245						25
f	241						33
g	482						95
h	260						25
i	490						160

* With webbing assembly fully extended.

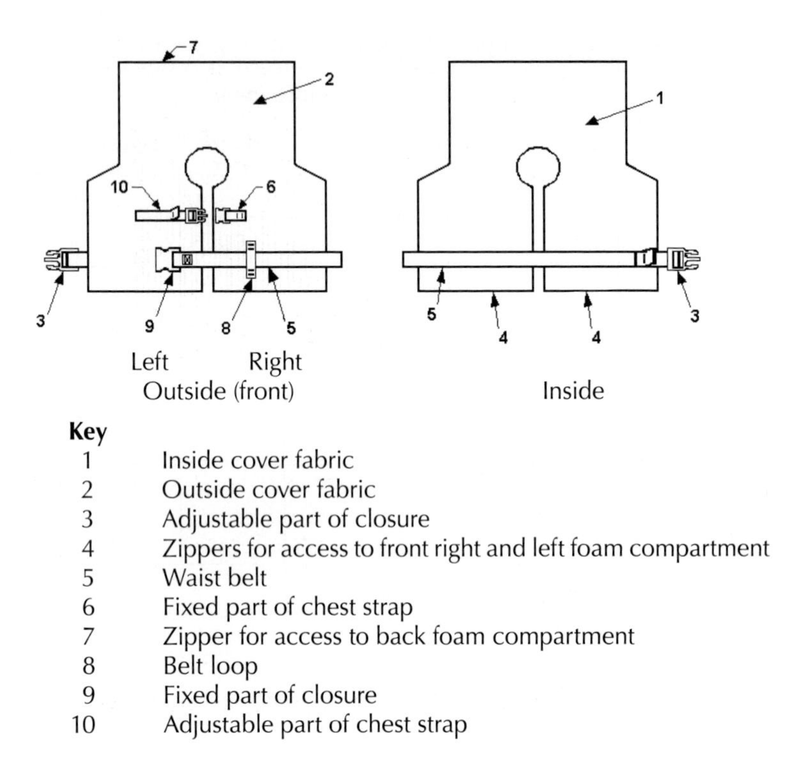

Key

1	Inside cover fabric
2	Outside cover fabric
3	Adjustable part of closure
4	Zippers for access to front right and left foam compartment
5	Waist belt
6	Fixed part of chest strap
7	Zipper for access to back foam compartment
8	Belt loop
9	Fixed part of closure
10	Adjustable part of chest strap

Figure C.1 – *General arrangement, right side out (outside and inside)*

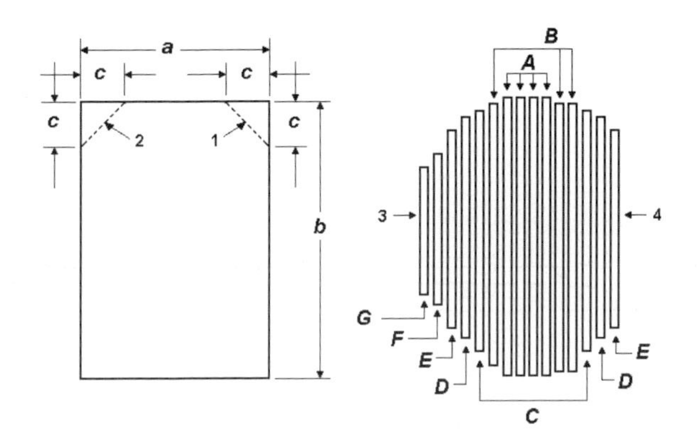

Key

1	Trim upper right corner only for left insert layers per table C.2
2	Trim upper left corner only for right insert layers per table C.2
3	Outside
4	Inside

Figure C.2 – *Front foam inserts (right and left sides)*

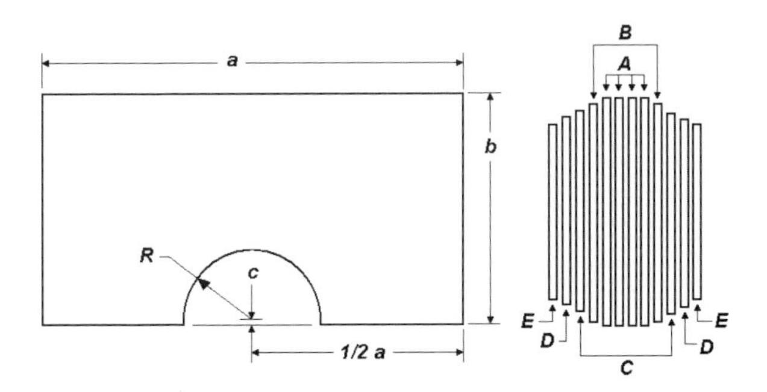

Figure C.3 – *Back foam insert*

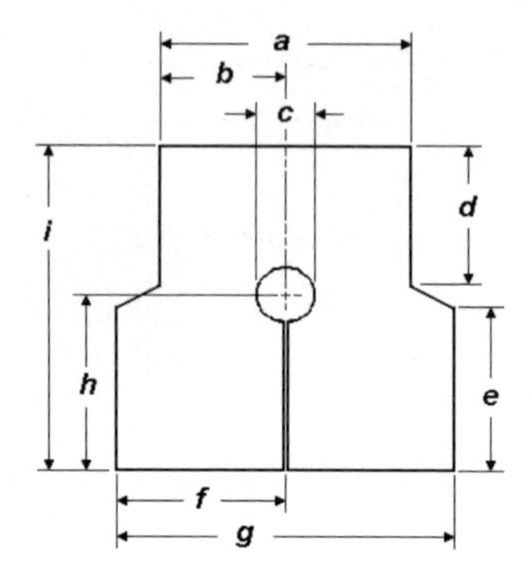

Figure C.4 – *Cover cut pattern (outside and inside covers)*

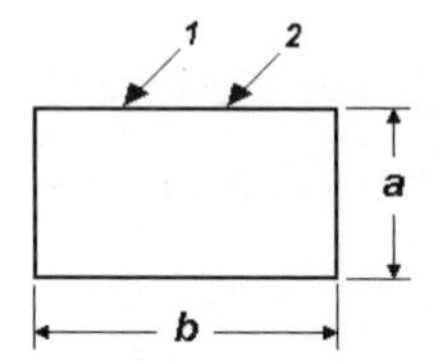

Key

1 Fabric reinforcements for chest strap attachments

2 Fabric reinforcement for waist belt and belt loop attachments

Figure C.5 – *Fabric reinforcements*

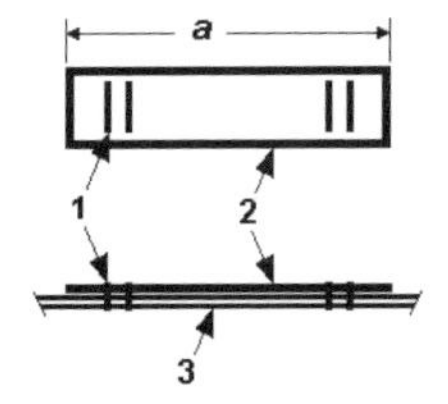

Key

1	Bar-tack
2	Webbing
3	Outer cover and reinforcement (shown on lower view only)

Figure C.6 – *Belt loop*

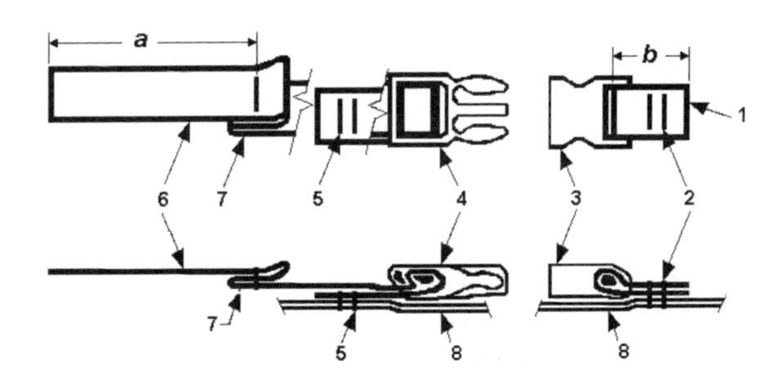

Key

1	Webbing
2	Double bar-tack (or Box-x) stitch
3	Fixed part of closure
4	Adjustable part of closure
5	Double bar-tack (or Box-x) stitch
6	Webbing
7	Tab
8	Outer cover and reinforcement (shown on lower view only)

Figure C.7 – *Chest strap assembly (adjustable part left and fixed part right)*

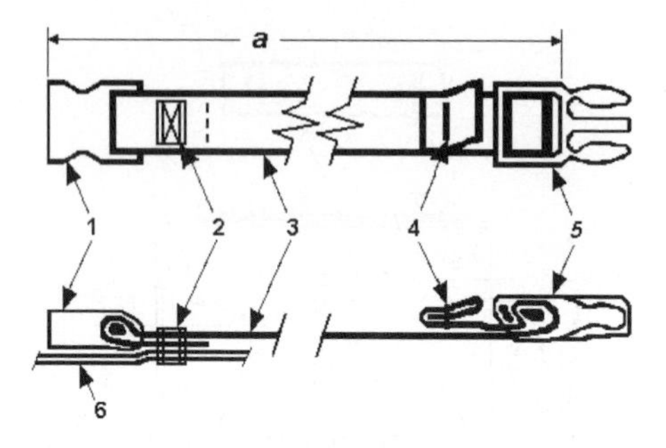

Key

1	Fixed part of closure
2	Box-x (or double bar-tack) stitch
3	Webbing
4	Tab, double fold webbing and secure with a bar-tack stitch
5	Adjustable part of closure
6	Outer left cover and reinforcement (shown on lower view only)

Figure C.8 – *Waist belt assembly*

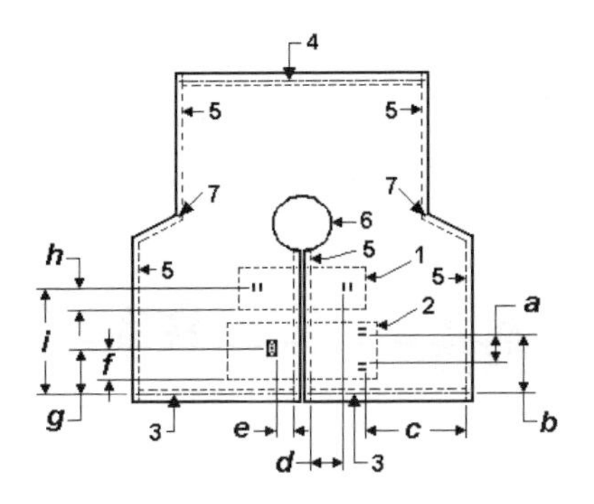

Key

1	Stitching on interior reinforcement for chest strap on right and left sides of outside cover only
2	Stitching on interior reinforcement for waist belt and belt loop on right and left sides of outside cover only
3	Fabric fold and zipper teeth line of engagement when attached to outside and inside covers
4	Fabric fold and zipper teeth line of engagement when attached to outside and inside covers
5	Lockstitch seams (with fabric face to face)
6	Lockstitch with 5 mm seam allowance and over-edge stitch (with fabric face to face)
7	After stitching seams cut relief

Figure C.9 – *Initial assembly (shown right side out, except as noted)*

Appendix

RTD Serial number: ________

Infant reference test device – buoyancy tracking and verification

To achieve repeatability in human subject and manikin testing, the overall buoyancy and distribution of buoyancy between the front and back of the RTD should be maintained within a tight tolerance as specified in table 1.

Table 1 – *SOLAS infant RTD buoyancy and tolerance*

Limit/Units	Front buoyancy[1, 2]	Back buoyancy[1]	Total buoyancy[1]	Buoyancy distribution[3]
Design/N	42	29	71	59.2% in front
Maximum/N	44.4	30.2	74.6	60.7% in front
Minimum/N	39.6	27.8	67.4	57.7% in front

[1] Values at or corrected to standard temperature and pressure.

[2] The left-to-right buoyancy distribution in the front inserts is to be within 1.3 N of each other.

[3] Buoyancy distribution is calculated by dividing the front buoyancy by the total buoyancy.

The buoyancy of a new RTD may exceed the allowable tolerance range until the normal shrinkage or compression of the foam inserts stabilizes. Until the buoyancies of the foam inserts have stabilized, buoyancy and distribution should be checked at regular intervals (perhaps weekly), and then at least monthly thereafter or whenever used for testing, whichever is longer (frequent use may require more frequent checks). Only RTDs with buoyancies within tolerance should be used for certification testing. A data sheet to document the buoyancy and buoyancy distribution of the RTD is attached.

To check buoyancy tolerances, foam inserts need to be removed from the device. Take care that all trapped air is removed when checking buoyancy and that layers are maintained in their proper sequence when reinstalled (considerable effort would be needed to remove entrapped air if testing the intact device).

Adjustment of buoyancy: At the time of manufacture the left-to-right distribution of buoyancy in the front inserts was adjusted to be within 1.3 N of each other. To achieve this tolerance, the layers were individually selected to achieve the cumulative insert buoyancy. If buoyancy of a new device exceeds the upper limits, one layer per compartment may be altered or replaced to bring the unit into compliance. The test house may need to add make-up layers (see figure 3) from time to time to maintain the front-to-back and side-to-side insert tolerances. If the front buoyancy is under the minimum value, measure the buoyancy of the right and left sides so that the proper distribution of buoyancy (no more than a 1.3 N difference) between the right and left front panels can be maintained.

Table 2 – *SOLAS infant RTD insert design buoyancies*

	Combined left front inserts (15 layers)	**Combined right front inserts (15 layers)**	**Combined back inserts (11 layers)**
Design (N)	21	21	29
S/N ______ Date:			

RTD BUOYANCY DATA SHEET

RTD Serial number/identification: ____________

Date	Left front buoyancy (N)[1]	Right front buoyancy (N)	Total front buoyancy (N)[2]	Total back buoyancy (N)	Total buoyancy (N)	Buoyancy distribution (% in front)	Remarks

[1] Left and right front buoyancy need not be checked if distribution is within tolerance.

[2] If the temperature and pressure at the time of measurement is not at standard conditions, these values should be corrected to standard temperature and pressure.

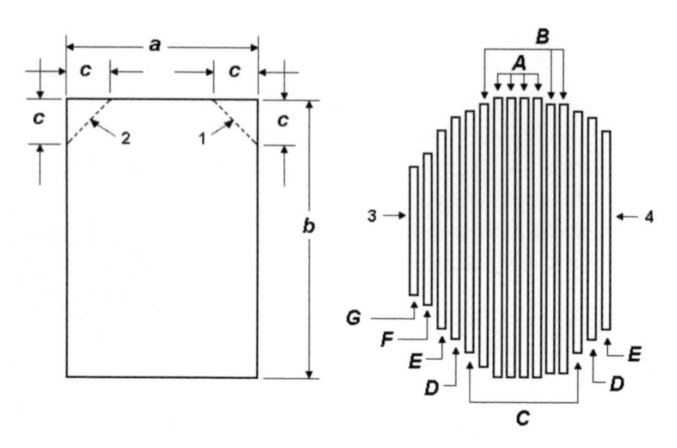

Key

1	Trim upper right corner only for left insert layers A to E
2	Trim upper left corner only for right insert layers A to E
3	Outside
4	Inside

Insert layer	Buoyancy (N)	Insert layer dimensions (mm)		
		a	*b*	*c*
A	1.7	140	190	28
B	1.6	133	184	28
C	1.4	127	178	28
D	1.3	120	172	28
E	1.1	108	165	28
F	1.0	95	160	0
G	0.8	83	140	0

Figure 1 – *Front foam insert specifications*

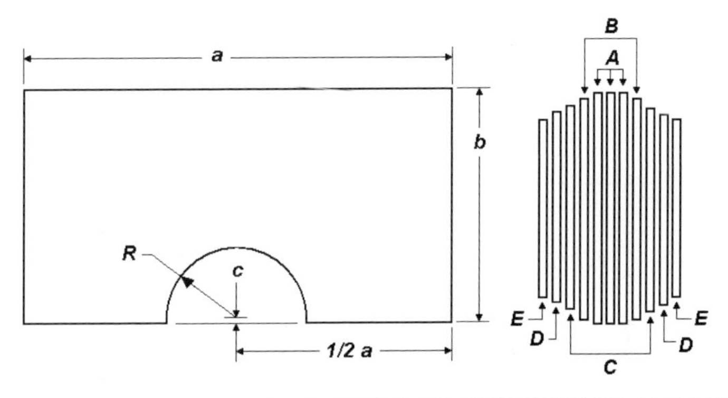

Insert layer	Buoyancy (N)	Insert layer dimensions (mm)			
		a	*b*	*c*	*R*
A	3.1	310	165	3	44
B	2.9	303	160	3	46
C	2.4	290	140	3	48
D	1.8	275	120	3	50
E	1.3	255	95	–3*	52

* Measured in direction opposite to that shown in figure.

Figure 2 – *Back foam insert specifications*

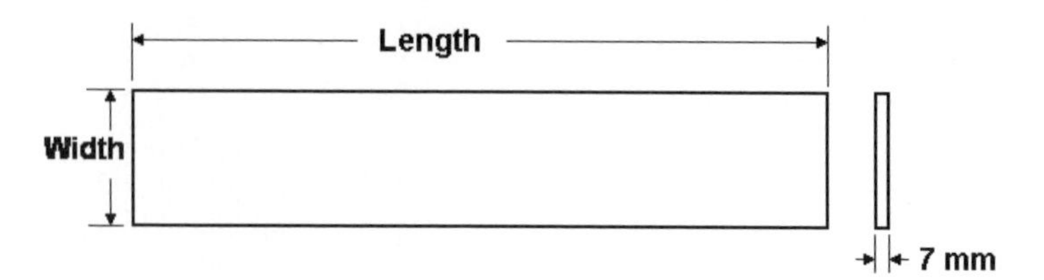

1 Any thickness of foam up to 7 mm is acceptable for a make-up layer.

2 For 7 mm thick foam, 15,300 mm^2 surface area equals approximately 1 N of buoyancy.

Make-up layer[1]	Buoyancy (approx.) (N)	Make-up layer dimensions (mm)	
		Length (mm)[2]	Width (mm)
Front	1	185	82
	1.5		123
Back	1	305	50
	1.5		75

[1] For 7 mm thick foam.

[2] The length for make-up layer is fixed to maintain proper placement within the lifejacket, but the width may vary to obtain desired buoyancy.

Figure 3 – *Infant RTD "make-up" foam insert sizes*

II

Code of practice for the evaluation, testing and acceptance of prototype novel life-saving appliances and arrangements

Resolution A.520(13)

(adopted on 17 November 1983)

THE ASSEMBLY,

RECALLING Article 16(j) of the Convention on the International Maritime Organization concerning the functions of the Assembly in relation to regulations and guidelines concerning maritime safety,

RECOGNIZING that prototype novel life-saving appliances and arrangements may be developed which do not fully meet the requirements of chapter III of the 1983 amendments to the International Convention for the Safety of Life at Sea, 1974, but which will provide the same or higher safety standards,

CONSIDERING the need to provide a code of practice for the evaluation, testing and acceptance of prototype novel life-saving appliances and arrangements to facilitate their acceptance by the Organization through amendment of chapter III of the International Convention for the Safety of Life at Sea, 1974,

NOTING regulation 4.3 of chapter III of the 1983 amendments to the International Convention for the Safety of Life at Sea, 1974,

HAVING CONSIDERED the recommendation made by the Maritime Safety Committee at its forty-eighth session,

1. ADOPTS the Code of Practice for the Evaluation, Testing and Acceptance of Prototype Novel Life-Saving Appliances and Arrangements, set out in the annex to the present resolution;

2. URGES Governments to ensure that prototype novel life-saving appliances and arrangements at least comply with the provisions of this Code of Practice;

3. REQUESTS the Maritime Safety Committee to keep this Code of Practice under review and to report as necessary to the Assembly.

Preamble

Chapter III of the International Convention for the Safety of Life at Sea, 1974 (hereinafter referred to as "the Convention"), was amended in 1983 by the Maritime Safety Committee by resolution MSC.6(48) to improve the standards of safety provided by life-saving appliances and arrangements on cargo ships and passenger ships, to make special provisions for certain types of cargo ships and to incorporate in the Convention life-saving appliances and arrangements which had been developed and accepted by Contracting Governments to the Convention (hereinafter referred to as "Contracting Governments") since 1974.*

It is recognized that the design and introduction of prototypes of improved novel life-saving appliances and arrangements is desirable and should be encouraged.

Prototype life-saving appliances and arrangements meeting the requirements of chapter III of the Convention, as amended, should be evaluated and tested in accordance with the Recommendation on Testing of Life-Saving Appliances.†

Prototype life-saving appliances and arrangements not meeting all the requirements of chapter III of the Convention, as amended, should be evaluated and tested to ensure that such appliances and arrangements at least comply with the provisions of this Code of Practice before being accepted by a Contracting Government. Subsequently, provisions for such life-saving appliances and arrangements should be incorporated in chapter III, through the amendment procedure prescribed in the Convention.

A Contracting Government proposing novel life-saving appliances and arrangements for adoption in accordance with the amendment procedure should submit full details of such appliances and arrangements and the results of prototype tests to the Organization to indicate that they comply with the requirements of this Code of Practice and in particular:

.1 provide all the functions of, and are equally effective as, the life-saving appliances and arrangements intended to be replaced; and

---

* Chapter III of the 1974 SOLAS Convention has since been completely revised (resolution MSC.47(66)). The new SOLAS chapter III entered into force on 1 July 1998.

† Resolution A.689(17) as amended by resolutions MSC.81(70), MSC.200(80), MSC.226(82) and MSC.274(85).

.2 do not affect the proper operation of any other life-saving appliance or arrangement installed on the ship.

Nothing in this Code shall be construed as derogating from or extending the rights of Administrations under regulation I/5 of the Convention to allow the fitting or carriage of any prototype novel life-saving appliance or arrangement.

1 General provisions

1.1 Purpose

This Code prescribes the appliance and arrangement criteria which should be taken into account and prototype tests which should be carried out for the evaluation of novel designs for international acceptance through the amendment procedure of the Convention.

1.2 Application

1.2.1 This Code applies to all prototype novel life-saving appliances and arrangements for which provisions have not been made in chapter III of the Convention, as amended, and for which Contracting Governments seek international acceptance.

1.2.2 Contracting Governments should apply the provisions of this Code when introducing equivalent life-saving appliances and arrangements pursuant to regulation I/5 of the Convention.

1.3 Definitions

1.3.1 The appropriate terms and definitions given in regulations I/2 and III/3 of the Convention, as amended, also apply to this Code. In addition, for the purpose of this Code unless expressly provided otherwise:

.1 *Equivalent* means a particular fitting, material, appliance, apparatus or arrangement, or any combination of fittings, materials, appliances, apparatus or arrangements allowed by an Administration to be fitted or carried in a ship in substitution for any of the requirements of chapter III of the Convention.

.2 *Prototype life-saving appliance or arrangement* is the first appliance or arrangement produced of a size and construction or performance characteristics differing from previous designs. If an appliance or arrangement is a modification of a previous design, only those characteristics affected by the modification are considered to be prototype characteristics which must be prototype tested.

- **.3** *Prototype tests* are those tests to which a prototype life-saving appliance or arrangement is subjected in order to test features which cannot be considered by the Administration on the basis of reasonable extrapolation of previous test results or experience.
- **.4** *Active survival craft* means a survival craft propelled by an engine.
- **.5** *Passive survival craft* means a survival craft which is not propelled by an engine.

2 General criteria

2.1 Operational readiness

2.1.1 The life-saving appliances on every ship should, either individually or collectively:

- **.1** be safely stowed and in a state of readiness for immediate use;
- **.2** provide means of abandonment of all persons on board in the shortest possible time; in the case of:
  - **.2.1** passenger ships, within a period of 30 min; and
  - **.2.2** any other ship, within a period of 10 min;
- **.3** include portable buoyancy equipment for the support and detection of persons in the water. Such equipment should be:
  - **.3.1** so distributed as to be readily available on both sides of the ship and as far as practicable on all open decks extending to the ship's side; at least one should be placed in the vicinity of the stern;
  - **.3.2** stowed in conspicuous places and so that they can be readily cast loose; they should not be permanently secured; and
  - **.3.3** fitted with suitable aids to assist detection;
- **.4** include sufficient individual buoyancy equipment for every person on board and, in addition, sufficient additional equipment to replace equipment which may become inaccessible;
- **.5** provide for the rescue and retrieval of persons in the water from survival craft or distressed ships;

.6 not be rendered inoperable by the effects of the marine atmosphere, seawater, fresh water, oil or fungus; in addition, where exposed to sunlight, they should be resistant to deterioration;

.7 not be damaged in stowage throughout an air temperature range from −30°C to +65°C and, if they are likely to be immersed in seawater during their use, be capable of operating throughout a seawater temperature range from −1°C to +30°C, unless other temperature ranges are relevant;

.8 on ships carrying hazardous cargo, provide protection for the crew from the effects of cargo hazards or fire during and after abandonment;

.9 where practicable, be constructed of fire-retardant materials; however, their attachments, fittings or equipment need not be of fire-retardant material provided they do not affect the efficient functioning of the appliance;

.10 be maintained and tested to ascertain that they meet the requirements of this Code.

2.1.2 Descriptions and instructions for operation, inspection, maintenance and functional testing should be provided for all the life-saving appliances, covering as appropriate the:

.1 purpose;

.2 operating description;

.3 physical description;

.4 operating instructions;

.5 requirements for inspection, maintenance, replacements and specialist servicing;

.6 requirements for operational testing, standards of performance, methods of adjustment; and

.7 fault-finding procedures.

2.1.3 Posters and signs in the vicinity of appliances and controls should:

.1 indicate the purpose of controls and procedures for operating the appliances or controls and give relevant instructions and warnings;

.2 be easily seen under emergency lighting conditions.

2.1.4 Instructions should be provided for each crew member which include the operations to be performed in relation to life-saving appliances in emergencies.

2.1.5 Spares and repair equipment should be provided for life-saving appliances or parts of life-saving appliances which are subject to excessive wear or consumption.

2.1.6 Life-saving appliances should be easy to inspect, maintain and test and, where applicable, be serviced at an approved servicing station.

2.1.7 Life-saving appliances should be simple to operate and should be so constructed that crew members can be easily familiarized with their use during practice musters and drills and require minimum prior training or experience.

2.1.8 Survival craft with launching arrangements should be stowed or located so that:

- **.1** neither the survival craft nor its stowage arrangements will interfere with the operation of any other survival craft or any other launching station;
- **.2** they are as near sea level as is safe and practicable and the embarkation position is at least 2 m above the waterline with the ship trimmed up to 10° and listed up to 20° either way, in the fully loaded condition, or to the angle at which the ship's weather deck becomes submerged, whichever is the least;
- **.3** boarding and launching can take place:
  - **.3.1** in cargo ships, directly from the stowed position and in the case of passive survival craft provided with launching devices, from a position immediately adjacent to the stowed position or from a position to which the survival craft is transferred prior to launching;
  - **.3.2** in passenger ships, either directly from the stowed position or from an embarkation deck but not both and, in the case of passive survival craft provided with launching devices, from a position immediately adjacent to the stowed position or from a position to which the survival craft is transferred prior to launching;
- **.4** they are kept in a state of continuous readiness and so that two crew members can carry out preparations for boarding and for launching in less than 5 min;
- **.5** after being prepared for boarding, they may be boarded in the case of:
  - **.5.1** passenger ships, rapidly;
  - **.5.2** cargo ships, in not more than 3 min;

.6 with the exception of that equipment which may be stowed in another location to protect it from pilferage or deterioration, they are fully equipped with all systems and components required for safe operation;

.7 when the survival craft are launched by falls or a fall and are not the additional survival craft provided on passenger ships, they are attached to their launching devices or within reach of the lifting hooks unless rapid and efficient means of transfer are provided which will not:

.7.1 be rendered inoperable under the conditions prescribed in paragraph 2.2.1 or by ship motion;

.7.2 reduce the time for preparing and boarding the survival craft prescribed in paragraphs 2.1.8.4 and 2.1.8.5;

.8 if arranged for throw-over launching, unless an adequate capacity of survival craft is available on both sides, the survival craft can readily be transferred for launching on either side of the ship;

.9 they are, as far as practicable, in a secure and sheltered position and protected from damage by fire and explosions.

2.1.9 Survival craft embarkation and launching arrangements should be provided except for those survival craft which are portable and are:

.1 boarded from a position on deck less than 4.5 m above the waterline in the lightest seagoing condition; or

.2 carried in excess of 200% of the number of persons on board.

2.1.10 Rescue craft should be stowed in such a way that:

.1 they are kept in a state of continuous readiness and can be launched within 5 min;

.2 neither the rescue craft nor its stowage arrangements will interfere with the operation of any survival craft at any other launching station.

2.1.11 Means for individual abandonment should:

.1 enable unassisted descent from deck to the water surface;

.2 be stowed in conspicuous and accessible locations ready for use:

.2.1 in the vicinity of survival craft launching areas; and

.2.2 in areas where persons may be isolated from survival craft due to fire or explosions.

2.1.12 Facilities should be provided for alerting all persons on board.

2.2 Abandonment

Abandonment should be possible:

.1 with the ship trimmed up to 10° and listed up to 20° either way or up to such angles of trim or list at which the ship's weather deck edge becomes submerged, whichever is the least, and on oil tankers, chemical tankers and gas carriers with a final angle of heel greater than 20° calculated in accordance with the International Convention for the Prevention of Pollution from Ships, 1973, as modified by the 1978 Protocol relating thereto and the recommendations of the Organization* as applicable, at the final angle of heel on the lower side;

.2 with the ship adrift in a seaway;

.3 in case of cargo ships of 20,000 gross tonnage and upwards, with the ship making way at speeds up to 5 knots in calm water;

.4 without depending upon any means other than gravity or stored power which is independent of the ship's power supplies to launch the survival craft.

2.3 Survival

2.3.1 Survival craft systems should:

.1 provide subsistence and protection for their complement under adverse weather conditions;

.2 have the capability of manoeuvring in a seaway.

2.3.2 Rescue craft should:

.1 provide protection for their complement under adverse weather conditions;

.2 have the capability of manoeuvring in a seaway.

2.4 Detection

2.4.1 Visual means of detection for survival craft should make it possible:

.1 for an aircraft at an altitude of up to 3,000 m to detect the survival craft at a range of at least 10 miles; and

---

* Refer to the damage stability requirements of the International Code for the Construction and Equipment of Ships Carrying Dangerous Chemicals in Bulk (IBC Code), as amended, adopted by the Maritime Safety Committee by resolution MSC.4(48) and the International Code for the Construction and Equipment of Ships Carrying Liquefied Gases in Bulk (IGC Code), as amended, adopted by the Maritime Safety Committee by resolution MSC.5(48).

.2 for a ship to detect the survival craft in a seaway in clear conditions at a range of at least 2 miles.

2.4.2 Visual means of detection for persons in the water should make it possible for a ship to detect the person in a seaway:

.1 in clear daytime conditions at a range of at least 0.2 miles;

.2 in clear night-time conditions at a range of at least 0.5 miles for a duration of at least 8 h.

2.5 Retrieval

2.5.1 Survival craft should:

.1 if passive, be capable of being towed at speeds of up to 3 knots;

.2 if active, be capable of being towed at speeds of up to 5 knots and be capable of towing other survival craft;

.3 permit a person to transfer from the survival craft in a seaway to a ship or helicopter.

2.5.2 Rescue craft should be capable of being towed at speeds of up to 5 knots and be capable of towing a survival craft.

2.5.3 Launching arrangements for rescue craft should provide safe launching from the ship in a seaway with the ship making way at speeds of up to 5 knots.

2.5.4 Retrieval arrangements for rescue craft should permit rapid recovery of the rescue craft with its rescue craft complement of at least six persons and equipment in a seaway.

3 Appliance criteria and testing of prototypes

3.1 Personal life-saving appliances

3.1.1 Buoyancy equipment intended to support and enable detection of persons in the water should:

.1 where required to be fitted with a buoyant lifeline, have a lifeline equal in length to at least twice the height at which it is stowed above the waterline in the lightest seagoing condition, or 30 m, whichever is the greater;

.2 be constructed to withstand a drop from the height at which it is stowed above the lightest seagoing waterline, or 30 m, whichever is the greater, without impairing its operating capability or that of its attached components;

.3 be capable of supporting not less than 14.5 kg of iron in fresh water for a period of 24 h;

.4 have means to enable persons to cling to the equipment;

.5 not sustain burning or continue melting after being totally enveloped in a fire for a period of 2 s;

.6 where required, be sufficiently heavy to operate release arrangements of the attached means of detection;

.7 be prototype tested with regard to paragraphs 3.1.1.2 to 3.1.1.6;

.8 where required, be provided with means of detection, complying with the requirements of paragraph 3.7.1.

3.1.2 Individual buoyancy equipment should:

.1 be so designed that after a demonstration a person can correctly don the equipment within a period of 1 min without assistance;

.2 not sustain burning or continue melting after being totally enveloped in a fire for a period of 2 s;

.3 be possible to wear without undue discomfort during abandonment and within a survival craft;

.4 if inflatable, inflate automatically upon immersion and be capable of being inflated manually and by mouth;

.5 if inflatable, perform effectively with any one buoyancy compartment inoperative;

.6 allow the wearer to jump into the water from a height of at least 4.5 m without sustaining injury and without dislodging or damaging the equipment;

.7 allow swimming and boarding of a survival craft in a seaway;

.8 in calm fresh water, be capable of lifting the mouth of a completely relaxed person wearing normal clothing at least 120 mm clear of the water;

.9 in calm fresh water, be capable of turning a completely relaxed person wearing normal clothing from any position in the water to one where the mouth is clear of the water within 5 s;

.10 have buoyancy which is not reduced by more than 5% after a 24 h period of submersion in fresh water;

.11 be prototype tested with regard to paragraphs 3.1.2.1 to 3.1.2.10 and with regard to paragraph 3.1.2.7 in a seaway;

.12 be provided with means of detection complying with paragraph 3.7.2; however, equipment provided on passenger ships on short international voyages need not comply with paragraph 3.7.2.2.

3.1.3 Individual garments for protection against hypothermia should:

.1 be so designed that after a demonstration a person can without assistance unpack and correctly don the garment and any required additional individual buoyancy equipment and clothing, within a period of 2 min, taking into account possible low ambient temperature conditions;

.2 not sustain burning or continue melting after being totally enveloped in a fire for a period of 2 s;

.3 not cause undue discomfort to the wearer during abandonment or in survival craft and permit the person wearing it and any additional individual buoyancy equipment and any associated clothing to:

.3.1 perform normal duties during abandonment;

.3.2 climb up and down a ladder at least 5 m in length;

.3.3 jump vertically into the water from a height of at least 4.5 m without sustaining injury, dislodging or causing damage to the garment or allowing undue ingress of water;

.4 allow a person wearing the garment and any required additional individual buoyancy equipment to swim and board a survival craft in a seaway;

.5 in calm fresh water, in conjunction with any required additional individual buoyancy equipment, be capable of lifting the mouth of a completely relaxed person wearing the garment at least 120 mm clear of the water;

.6 in calm fresh water permit a person wearing the garment and any required additional individual buoyancy equipment to turn from any position to one where the mouth is clear of the water in not more than 5 s;

.7 if provided with buoyancy, not suffer a loss of buoyancy of more than 5% after a 24 h period of submersion in fresh water;

.8 be prototype tested with regard to paragraphs 3.1.3.1 to 3.1.3.7 and with regard to paragraph 3.1.3.4 in a seaway;

.9 if meant to be worn without additional buoyancy equipment or on top of such equipment, to be provided with means of detection complying with paragraph 3.7.2.

3.1.4 In addition to meeting the requirements of paragraph 3.1.3, an individual garment for long-term immersion should:

.1 when used over light clothing and with any required additional individual buoyancy equipment, allow the wearer following one jump into the water from a height of 4.5 m to float in calm circulating water of between 0°C and 2°C for a period of 6 h during which period the wearer's body core temperature should not fall more than 2°C;

.2 permit the wearer, on completion of the above test, to be able to pick up a pencil and write;

.3 be prototype tested with regard to paragraphs 3.1.4.1 and 3.1.4.2.

3.1.5 In addition to meeting the requirements of paragraph 3.1.3, an individual garment for short-term immersion should:

.1 when worn in conjunction with warm clothing and any required additional individual buoyancy equipment, following one jump by the wearer into the water from a height of 4.5 m, continue to provide sufficient thermal protection to ensure that when worn for a period of 1 h in calm circulating water at a temperature of 5°C the wearer's body core temperature should not fall more than 2°C;

.2 permit the wearer, on completion of the above test, to be able to pick up a pencil and write;

.3 be prototype tested with regard to paragraphs 3.1.5.1 and 3.1.5.2.

3.2 Survival craft

3.2.1 Survival craft should:

.1 where arranged to be suspended by falls and lowered to the water by means of a launching device, be of sufficient strength to withstand:

.1.1 if rigid, an overload of 100% of the total mass of the survival craft when loaded with its full complement of persons and equipment, without suffering significant residual deflection on removal of that load, except that in the case of a survival craft constructed of metal, the overload should be 25%;

.1.2 if inflatable, a load of four times the mass of its full complement and equipment at an ambient temperature of +20°C without pressure relief of inflated compartments, and a load of 1.1 times the mass of its full complement and equipment at an ambient temperature of –30°C;

.1.3 when fully loaded without sustaining damage that would affect its efficient functioning:

- a sideways impact against a rigid vertical surface at an impact velocity of at least 3.5 m/s; and
- a drop into the water from a height of 3 m;

.2 in the case of a self-righting partially enclosed and totally enclosed survival craft, protect its occupants when subjected to the sideways impact referred to in paragraph 3.2.1.1.3;

.3 if inflatable:

.3.1 withstand an inflation test pressure of at least 3 times the working pressure and be so arranged that the pressure cannot exceed twice the working pressure; and

.3.2 inflate with a non-toxic gas within a period of 1 min at an ambient temperature between 18°C and 20°C and within a period of 3 min at an ambient temperature of –30°C;

.4 where arranged for free-fall launching, have sufficient strength and diving characteristics to withstand a fall into the sea from the maximum height at which it is designed to be stowed, taking into account unfavourable conditions of trim up to 10°

and a list up to 20° either way, without impairing its operating capabilities or causing injury to its occupants;

.5 where required to float free, be stowed in such a manner as to permit it to float free from its stowage and break free from the ship in an operational condition when the ship sinks;

.6 if inflatable, withstand repeated jumps on to it from a height of at least 4.5 m above the water;

.7 be approved for the maximum number of persons it is permitted to accommodate, as decided by practical seating tests afloat and based upon the number of adult persons wearing individual buoyancy equipment who can be seated without in any way interfering with the normal operation of its equipment or means of propulsion;

.8 when prepared for launching, permits its full complement of persons excluding any stretcher cases to board rapidly and in the case of cargo ships in not more than 3 min from the time the instruction to board is given;

.9 permit embarkation of stretcher cases;

.10 have arrangements to secure it to the ship by a painter of adequate strength and of a length equal to at least twice the distance from its stowed position to the lightest seagoing waterline of 15 m, whichever is the greater;

.11 in the case of a self-righting partially enclosed or totally enclosed survival craft, unless capable of operating safely in the upside-down position, have such strength and stability that it is inherently or automatically self-righting when all entrances and openings are closed watertight, all equipment is secured and the full complement of persons are secured to their seats with safety belts;

.12 in the case of a passive survival craft, unless capable of operating safely in the upside-down position, have such sufficient strength and stability that:

.12.1 it is self-righting; or

.12.2 it can be readily righted, in a seaway, by one person unassisted;

.13 when fully or partly loaded maintain its operational effectiveness when drifting in a seaway;

.14 have sufficient buoyancy to support its full complement even when holed in any one location below the waterline without loss of buoyancy material or other damage, to automatically or inherently attain a position which provides an above-water escape for its occupants;

.15 have a freeboard, measured from the waterline to the lowest opening through which the survival craft may become flooded, of not less than 1.5% of the survival craft's length when loaded with one half of its full complement seated to one side of the centreline;

.16 be provided with effective means for bailing or be self-bailing in a seaway, but self-righting partially enclosed survival craft should be automatically self-bailing in a seaway;

.17 provide protection for its complement against wind, rain and spray, adequate ventilation and protection for its complement at all ambient temperatures between –15°C and +30°C;

.18 be designed with due regard to the safety of persons on board with regard to slippery or hot surfaces and sharp edges;

.19 be possible to manoeuvre;

.20 provide means for persons in the water to cling to the survival craft;

.21 permit persons to board the survival craft from the water when wearing individual buoyancy equipment;

.22 permit those on board the survival craft, when wearing individual buoyancy equipment, to recover persons from the water without their assistance;

.23 be provided with manually controlled lighting sufficient to permit reading of instruction material and to facilitate operations at night with a power capacity sufficient for at least 12 h;

.24 carry provisions, water and equipment for the full complement;

.25 be of sufficient strength and have sufficiently strong fixtures and painters to be:

- **.25.1** towed at speeds up to 3 knots in the case of passive survival craft;
- **.25.2** towed at all speeds up to 5 knots in the case of active survival craft;

.25.3 launched with the ship making headway at speeds up to 5 knots in the case of active survival craft intended for use on cargo ships of 20,000 gross tonnage and upwards;

.26 have means of permitting watertight restowage, where equipment is required to be stowed in watertight containers;

.27 be so arranged that in a seaway, a person in the survival craft may be picked up by helicopter or transferred to a ship by ladder or net without assistance from other persons in the survival craft;

.28 be prototype tested with regard to paragraphs 3.2.1.1 to 3.2.1.6, 3.2.1.8 to 3.2.1.12, 3.2.1.14, 3.2.1.15, 3.2.1.17, 3.2.1.19 to 3.2.1.23 and 3.2.1.25 and with regard to paragraphs 3.2.1.7 and 3.2.1.16 in a seaway;

.29 be provided with means for location and detection complying with paragraph 3.7.4.

3.2.2 In addition to meeting the requirements of paragraph 3.2.1, active survival craft should:

.1 have a means of propulsion, capable of being started manually or by two independent power sources and operated at an ambient temperature of –15°C within 2 min of commencing the engine start procedure using if necessary starting aids, unless, having regard to the particular voyages in which the ship carrying the craft is constantly engaged, another minimum starting and operating temperature is appropriate;

.2 if self-righting partially enclosed or totally enclosed, have a means of propulsion capable of running in an inverted position during capsize of the survival craft and of continuing to run after returning to the upright position unless it is stopped automatically when inverted and is easily restarted after the survival craft has returned to the upright position and, in the case of a self-righting partially enclosed survival craft, the water has drained from the survival craft. Capsizing should not cause a significant spill of oil into the survival craft;

.3 have sufficient mobility and manoeuvrability in a seaway to allow retrieval of persons from the water, marshalling of passive survival craft and to allow the largest passive survival craft carried on the ship to be towed at a speed of 2 knots in calm water;

.4 be capable of maintaining a speed of at least 6 knots for at least 24 h;

.5 be designed with due regard to the safety of persons in the water and the possibility of damage to the propulsion system by floating debris;

.6 be prototype tested with regard to paragraphs 3.2.2.1, 3.2.2.2 and 3.2.2.4 and with regard to paragraphs 3.2.2.3 in a seaway.

3.2.3 In addition to meeting the requirements of paragraphs 3.2.1 and 3.2.2, a survival craft with a self-contained air support system should:

.1 be capable of proceeding at maximum power for at least 10 min without exposing its complement to harmful gases or creating a subatmospheric pressure within the survival craft;

.2 be prototype tested with regard to paragraph 3.2.3.1 by running the survival craft for at least 10 min while maintaining an overpressure within the survival craft of not more than 20 mbar.

3.2.4 In addition to meeting the requirements of paragraphs 3.2.1, 3.2.2 and 3.2.3, fire-protected survival craft should be prototype tested by enveloping the survival craft, with its means of propulsion running, in a fire for a period of at least 8 min during which time there should be no ingress of harmful fumes and habitable temperatures should be maintained within the survival craft.

3.3 Rescue craft

Rescue craft should comply with the provisions of paragraphs 3.2.1.1.1, 3.2.1.4, 3.2.1.11, 3.2.1.13, 3.2.1.15 to 3.2.1.18, 3.2.1.20 to 3.2.1.22, 3.2.1.25.2, 3.2.1.29, 3.2.2.1, 3.2.2.3 and 3.2.2.5, and in addition should:

.1 where arranged to be suspended by a fall or falls and lowered to the water by means of a launching device, be of sufficient strength and fendered to withstand:

.1.1 if inflated, a load of 4 times the mass of the rescue boat and its full complement of persons and equipment at an ambient temperature of 20°C with all relief valves inoperative and 1.1 times the mass of the rescue boat and its full complement of persons and equipment at an ambient temperature of –30°C with all relief valves operative;

.1.2 when fully loaded, without sustaining damage that would affect its efficient functioning:

– a sideways impact against a rigid vertical surface with an impact velocity of at least 3.5 m/s; and

– a drop into the water from a height of 3 m;

.2 be capable of carrying at least five persons seated and at least one person lying down;

.3 be approved for the maximum number of persons to be decided by practical seating tests in a seaway and based upon the number of adult persons wearing individual buoyancy equipment who can be seated without in any way interfering with the normal operation of its equipment or means of propulsion;

.4 when preparing for launching, permit its full complement of persons, excluding any stretcher cases, to board in not more than 3 min;

.5 have sufficient buoyancy and stability to support its full complement even when holed in any one location and open to the sea;

.6 offer protection against exposure to sea spray;

.7 be capable of maintaining a speed of at least 6 knots for at least 4 h in a seaway;

.8 be prototype tested with regard to paragraphs 3.2.1.1.1, 3.2.1.4, 3.2.1.16, 3.2.1.22, 3.2.2.1, 3.2.2.3, 3.3.1.1, 3.3.1.2, 3.3.2 to 3.3.4, 3.3.6 and 3.3.7 and with regard to paragraphs 3.2.1.11, 3.2.1.13, 3.2.1.18, 3.2.1.20, 3.2.1.25.2, 3.2.1.29, and 3.3.5 in a seaway.

3.4 Launching arrangements

3.4.1 Launching arrangements for survival craft should:

.1 provide safe launching under normal conditions and with the ship trimmed up to 10° and listed up to 20° either way or up to such angles of trim or list at which the ship's weather deck edge becomes submerged, whichever is the least, and on oil tankers, chemical tankers and gas carriers with a final angle of heel greater than 20° calculated in accordance with the International Convention for the Prevention of Pollution from Ships, 1973, as modified by the 1978 Protocol relating thereto, and the recommendations of the Organization* as applicable, at the final angle of heel on the lower side;

.2 in the case of cargo ships of 20,000 gross tonnage and upwards, provide safe launching with the ship making headway at speeds of up to 5 knots;

.3 where they include falls and a winch, ensure that the speed at which the survival craft is lowered into the water is controlled by suitable means and such that the rate of descent after initial acceleration is at least a rate obtained from the formula:

$$S = 0.4 + 0.02H$$

where:

S = speed of lowering in metres per second and

H = height in metres from the uppermost point of suspension to the waterline in the lightest seagoing condition;

.4 meet the requirements of the Administration for the maximum lowering speed of the survival craft to ensure the protection of its occupants from excessive forces and to prevent damage to the launching arrangements taking into account inertial forces during an emergency stop;

.5 except for winch brakes, be of sufficient strength to withstand a static load of at least 2.2 times their maximum working load;

.6 where they include winch brakes, have winch brakes of sufficient strength to withstand:

.6.1 a static test with a proof load of 1.5 times the maximum working load;

.6.2 a dynamic test with a proof load of not less than 1.1 times the maximum working load at maximum lowering speed;

.7 be capable of being actuated by one person from a position on board the survival craft and from a position on deck from which the launching can be observed and the release of the

* Refer to the damage stability requirements of the International Code for the Construction and Equipment of Ships Carrying Dangerous Chemicals in Bulk (IBC Code), as amended, adopted by the Maritime Safety Committee by resolution MSC.4(48) and the International Code for the Construction and Equipment of Ships Carrying Liquefied Gases in Bulk (IGC Code), as amended, adopted by the Maritime Safety Committee by resolution MSC.5(48).

survival craft from the launching arrangements should be possible from a position on board the survival craft;

.8 if based on launching by a fall or falls, have a release mechanism which will release the survival craft into the water without causing damage to the survival craft;

.9 if for float-free release:

.9.1 release the survival craft from its stowed position;

.9.2 be designed to minimize the possibility of the survival craft becoming fouled in davits, cranes, rigging or superstructure;

.9.3 not be released unintentionally by such forces as water on deck and heavy weather;

.9.4 not be affected by shipboard vibration;

.9.5 provide for manual release;

.9.6 of survival craft having a rigid enclosure, release and launch the survival craft in all conditions of loading without it becoming swamped and should in addition allow the release and launch of the survival craft from the operator's position within the survival craft;

.10 if arranged for free-fall launching, ensure that the survival craft is released clear of the ship;

.11 ensure that the survival craft is upright in the water after launching into a seaway, unless the survival craft is self-righting, may be used in the upside-down condition or can be righted by one person in the water;

.12 be prototype tested with regard to paragraphs 3.4.1.2 to 3.4.1.11 and with regard to paragraph 3.4.1.1 in a seaway.

3.4.2 Launching arrangements for rescue craft should comply with the provisions of paragraphs 3.4.1.3 to 3.4.1.7, 3.4.1.10 and in addition should:

.1 provide safe launching when the ship is:

.1.1 in a seaway; and

.1.2 making headway at speeds of up to 5 knots;

.2 if based on launching by a fall or falls, have a release mechanism which will release the rescue craft into the water without causing damage to the rescue craft or injury to its complement; and

.3 be prototype tested with regard to paragraphs 3.4.1.3 to 3.4.1.7, 3.4.1.10, 3.4.2.1 and 3.4.2.2 and with regard to paragraph 3.4.2.1.1 in a seaway.

3.5 Retrieval arrangements

3.5.1 Retrieval arrangements for active survival craft should:

.1 provide for safe retrieval of the survival craft in a seaway;

.2 return the survival craft to its position of stowage and readiness for use;

.3 be of sufficient strength to withstand a static load of at least 2.2 times its working load except for winch brakes which should withstand a static load of 1.5 times the maximum working load;

.4 be prototype tested with regard to paragraphs 3.5.1.2 and 3.5.1.3 and with regard to paragraph 3.5.1.1 in a seaway.

3.5.2 Retrieval arrangements for rescue craft should:

.1 provide for safe retrieval of the craft in a seaway;

.2 provide a retrieval speed of at least 0.3 m/s when loaded with its rescue craft complement of at least six persons and equipment;

.3 return the craft to its position of stowage and readiness for use;

.4 be of sufficient strength to withstand a static load of at least 2.2 times its working load except for the winch brakes which should withstand a static load of 1.5 times the maximum working load;

.5 be prototype tested with regard to paragraphs 3.5.2.2 to 3.5.2.4, and with regard to paragraph 3.5.2.1 in a seaway.

3.6 Means of passing a line

3.6.1 Means of passing a line from the ship should:

.1 be capable of throwing a line with reasonable accuracy over a distance of at least 230 m; and

.2 be prototype tested with regard to paragraph 3.6.1.1.

3.7 Communications (alerting and detection)

3.7.1 Buoyancy equipment intended to support and enable detection of persons in the water should:

.1 if required, have active means of detection attached which is automatically activated when the buoyancy equipment is released and makes it possible to detect the buoyancy equipment in a seaway visually from a ship at a range of at least 1 mile for a period of:

.1.1 at least 15 min under clear daytime conditions; and

.1.2 at least 2 h under clear night-time conditions;

and when carried on tankers, such active means of detection should be of a type which cannot cause ignition of flammable vapours;

.2 have passive means of detection which makes it possible to detect the buoyancy equipment in a seaway visually from a ship at a range of at least 0.3 miles under clear daytime conditions and, when illuminated by a searchlight, from a range of at least 0.3 miles under clear night-time conditions;

.3 identify the ship on which it is carried;

.4 be prototype tested with regard to paragraph 3.7.1.1 and with regard to paragraphs 3.7.1.1.1, 3.7.1.1.2 and 3.7.1.2 in a seaway.

3.7.2 Individual buoyancy equipment and garments for protection against hypothermia should:

.1 have a manually controlled active means of detection which makes it possible to detect a person in a seaway audibly at a range of at least 0.2 miles in calm weather;

.2 have active means of detection which makes it possible to detect a person in a seaway visually at a range of at least 0.5 miles under clear night-time conditions for a period of not less than 8 h;

.3 have passive means of detection which makes it possible to detect a person in a seaway visually from a ship at a range of at least 0.2 miles under clear daytime conditions and, when illuminated by a searchlight, from a range of at least 0.2 miles under clear night-time conditions; and

.4 be prototype tested with regard to paragraph 3.7.2.1 and with regard to paragraphs 3.7.2.2 and 3.7.2.3 in a seaway.

3.7.3 Survival craft should:

.1 have active means of detection which makes it possible to visually locate or detect the survival craft in a seaway from a ship or an aircraft, as appropriate:

.1.1 at an altitude of 3,000 m at a range of at least 10 miles under clear daytime and night-time conditions for a period of at least 40 s;

.1.2 at a range of at least 3 miles under clear night-time conditions for a period of at least 1 min;

.1.3 at a range of at least 2 miles under clear daytime conditions for a period of at least 3 min;

.1.4 at a range of at least 2 miles under clear night-time conditions, which means should be manually operated, have sufficient capacity for at least 12 h operation and, in the case of passive survival craft, should be automatically activated when launched;

.2 have passive means of detection which makes it possible to locate and detect the survival craft in a seaway visually from a ship at a range of at least 1 mile in clear daytime conditions and, when illuminated by a searchlight, under clear night-time conditions;

.3 identify the ship on which they are carried;

.4 be provided with items that are prototype tested with regard to paragraphs 3.7.3.1 and 3.7.3.2 in a seaway.

3.7.4 Rescue craft should:

.1 have active means of detection which makes it possible to detect the rescue craft in a seaway from the ship on which they are carried, visually at a range of at least 2 miles under clear daytime and night-time conditions;

.2 have passive means of detection which makes is possible to detect the rescue craft in a seaway visually from the ship on which they are carried at a range of at least 1 mile under clear daytime conditions and, when illuminated by a searchlight, under clear night-time conditions;

.3 identify the ship on which they are carried;

.4 be provided with items that are prototype tested with regard to paragraphs 3.7.4.1 and 3.7.4.2 in a seaway.

3.7.5 The ship should be provided with active means of detection which makes it possible to detect and locate the ship from an altitude of at least 3,000 m at a range of at least 10 miles under clear daytime and night-time conditions for a period of at least 40 s.